*Muirhead Library of Philosophy*

# PERCEPTION

MUIRHEAD

*Muirhead Library of Philosophy*

## EPISTEMOLOGY
## In 6 Volumes

| | | |
|---|---|---|
| I | The Ways of Reason | *LaLumia* |
| II | The Nature of Physical Existence | *Leclerc* |
| III | Perception | *Locke* |
| IV | Identity & Reality | *Meyerson* |
| V | Belief | *Price* |
| VI | Reason and Scepticism | *Slote* |

# PERCEPTION

# And Our Knowledge of the External World

DON LOCKE

LONDON AND NEW YORK

First published in 1967 by Routledge

Published 2013 by Routledge
2 Park Square, Milton Park, Abingdon, Oxfordshire OX14 4RN
711 Third Avenue, New York, NY 10017, USA

First issued in paperback 2014

*Routledge is an imprint of the Taylor & Francis Group, an informa business*

© 1967 Taylor & Francis

All rights reserved. No part of this book may be reprinted or reproduced or utilized in any form or by any electronic, mechanical, or other means, now known or hereafter invented, including photocopying and recording, or in any information storage or retrieval system, without permission in writing from the publishers.

The publishers have made every effort to contact authors/copyright holders of the works reprinted in the *Muirhead Library of Philosophy*. This has not been possible in every case, however, and we would welcome correspondence from those individuals/companies we have been unable to trace.

These reprints are taken from original copies of each book. In many cases the condition of these originals is not perfect. The publisher has gone to great lengths to ensure the quality of these reprints, but wishes to point out that certain characteristics of the original copies will, of necessity, be apparent in reprints thereof.

*British Library Cataloguing in Publication Data*
A CIP catalogue record for this book
is available from the British Library

Perception
ISBN 978-0-415-29562-8
Epistemology: 6 Volumes
ISBN 0-415-29528-9
Muirhead Library of Philosophy: 95 Volumes
ISBN 0-415-27897-X

ISBN: 978-1-138-87078-9 (pbk)
ISBN: 978-0-415-29562-8 (hbk)

# MUIRHEAD LIBRARY OF PHILOSOPHY

An admirable statement of the aims of the Library of Philosophy was provided by the first editor, the late Professor J. H. Muirhead, in his description of the original programme printed in Erdmann's *History of Philosophy* under the date 1890. This was slightly modified in subsequent volumes to take the form of the following statement:

'The Muirhead Library of Philosophy was designed as a contribution to the History of Modern Philosophy under the heads: first of Different Schools of Thought—Sensationalist, Realist, Idealist, Intuitivist; secondly of different Subjects—Psychology, Ethics, Aesthetics, Political Philosophy Theology. While much had been done in England in tracing the course of evolution in nature, history, economics, morals and religion, little had been done in tracing the development of thought on these subjects. Yet "the evolution of opinion is part of the whole evolution".'

'By the co-operation of different writers in carrying out this plan it was hoped that a thoroughness and completeness of treatment, otherwise unattainable, might be secured. It was believed also that from writers mainly British and American fuller consideration of English Philosophy than it had hitherto received might be looked for. In the earlier series of books containing, among others, Bosanquet's *History of Aesthetic*, Pfleiderer's *Rational Theory since Kant*, Albee's *History of English Utilitarianism*, Binar's *Philosophy and Political Economy*, Brett's *History of Psychology*, Ritchie's *Natural Rights*, these objects were to a large extent effected.

'In the meantime original work of a high order was being produced both in England and America by such writers as Bradley, Stout, Bertrand Russell, Baldwin, Urban, Montague, and others, and a new interest in foreign works, German, French and Italian, which had either become classical or were attracting public attention, had developed. The scope of the Library thus became extended into something more international, and it is entering on the fifth decade of its existence in the hope that it may contribute to that mutual understanding between countries which is so pressing a need of the present time.'

The need which Professor Muirhead stressed is no less pressing today, and few will deny that philosophy has much to do with enabling us to meet it, although no one, least of all Muirhead himself, would regard that as the sole, or even the main, object of philosophy. As Professor Muirhead continues to lend the distinction of his name to the

Library of Philosophy it seemed not inappropriate to allow him to recall us to these aims in his own words. The emphasis on the history of thought also seemed to me very timely; and the number of important works promised for the Library in the near future augur well for the continued fulfilment, in this and other ways, of the expectations of the original editor.

H. D. LEWIS

I would like to thank Professor A. R. White and Professor K. W. Britton for helpful comments on earlier drafts of this book. They are not responsible for the faults that remain.

# PERCEPTION
AND OUR KNOWLEDGE OF THE EXTERNAL WORLD

BY

DON LOCKE

LONDON: GEORGE ALLEN & UNWIN LTD
NEW YORK: HUMANITIES PRESS INC

FIRST PUBLISHED IN 1967

*This book is copyright under the Berne Convention. Apart from any fair dealing for the purpose of private study, research, criticism or review, as permitted under the Copyright Act, 1965, no portion may be reproduced by any process without written permission. Enquiries should be addressed to the publisher*

© *George Allen and Unwin Ltd, 1967*

*Library of Congress Catalog Card Number: 66.28126*

PRINTED IN GREAT BRITAIN
*in 11 on 12 point Imprint type*
BY UNWIN BROTHERS LTD
WOKING AND LONDON

*To Margaret, Nicola and Stephen*

# CONTENTS

1. INTRODUCTION
   1. The Problems of Perception         page 13
   2. Some Crucial Definitions                15

   PART ONE: PERCEPTION

2. AN ANALYSIS OF PERCEPTION
   1. Sensory Awareness                       27
   2. Noticing                                30
   3. Perception-that                         32

3. THE PHILOSOPHICAL THEORIES OF PERCEPTION
   1. The Three Theories                      35
   2. The Introduction of Percepts            37
   3. The Theories as Alternative Languages   40

4. PHENOMENALISM
   1. Phenomenalism, Idealism and Realism     47
   2. The Truth in Phenomenalism              50
   3. Objections to Phenomenalism             54
   4. 'Pragmatic' Phenomenalism               66

5. THE ARGUMENTS TO SENSE-DEPENDENCE
   1. Primary Qualities and Secondary Qualities  68
   2. The Argument from Sensations            78
   3. Bodily Sensations                       83

6. THE ARGUMENT FROM ILLUSION
   1. Non-veridical Perception                92
   2. Reality                                 95
   3. Appearance                             103
   4. Illusion                               105
   5. Three Arguments                        108

7. THE ARGUMENTS FROM SCIENCE
   1. The Causal Argument                    113
   2. The Argument from Physics              118
   3. The Nature of External Reality         120

8. THE DEFENCE OF REALISM
   1. The Epistemological Arguments          126
   2. Moore's Proof of an External World     129
   3. The Argument from the Conceptual Scheme 131

## 9. THE CHOICE BETWEEN THE THEORIES
1. The Theories Considered — 135
2. The Sensibilia Theory — 138

## PART TWO: OUR KNOWLEDGE OF THE EXTERNAL WORLD

### 10. PERCEPTION AND KNOWLEDGE
1. The Problem — 145
2. Knowing and Saying 'I know' — 145
3. The Right to be Sure — 148
4. An Analysis of Knowledge — 150
5. Scepticism — 152
6. Empiricism — 157
7. The Foundations of Empirical Knowledge — 161

### 11. SENSE DATA
1. The Need for a Sense Datum Terminology — 164
2. Moore's Methods of Explaining Sense Data — 168
3. Immediate Perception — 172
4. The Nature of Sense Data — 178
5. Sense Datum Knowledge — 184
6. Sense Data and What We Take Ourselves to Perceive — 192
7. Sense Data and Appearances — 197
8. The Importance of Sense Data — 201

### 12. OUR KNOWLEDGE OF EXTERNAL EXISTENCE
1. The Structure of Sense Knowledge — 204
2. Reductionism and Constructivism — 210
3. Derivations — 215
4. A Derivation of Sense Datum Knowledge — 217
5. A Derivation of Knowledge of Reality — 222
6. The Epistemological Moral — 228
7. Our Knowledge as Agents — 230

LIST OF WORKS CITED — 237

INDEX — 241

# CHAPTER 1
# INTRODUCTION

## 1.1 THE PROBLEMS OF PERCEPTION

Why should philosophers be interested in perception at all? To the layman perception seems to be a problem for psychologists or physiologists, or else no problem at all. And some philosophers are inclined to agree. Yet I think there are two important routes by which a philosopher will be led to discuss perception. One runs via the Philosophy of Mind, and the other via Epistemology of the Theory of Knowledge.

Philosophy is mainly concerned with the examination and clarification of concepts, and since Perception is a concept like any other, we can naturally expect such questions as 'What is it to perceive something?' or 'What do we mean when we say that something is perceived?' Yet it is hard to see why these questions should be specially important, why they in particular should attract so much attention. One reason is that the philosopher has a special interest in the mind and mental phenomena, and in so far as perception is some sort of mental process, so far he will be particularly interested in perception. But more important still is the fact that many philosophers have held that to perceive something is not to be aware of the sorts of things we naturally think of ourselves as perceiving—houses and horses, babies and bottles—but to be aware of certain mental entities, variously termed ideas, sensations, sense-impressions, sense data, percepts, etc., which exist not in the world around us but, in some sense, in our minds. If any such theory is correct, then the Philosophy of Mind has the important and difficult task of examining these entities, their nature, their status, their existence, their location, their role in the functions and constitution of the mind, and so on.[1] In short, philosophers would not be so interested in the question 'What is it to perceive something?' if other philosophers, and some scientists, had not answered it in some strange ways.

It has been suggested that philosophical theories of perception are *really* theories about what we mean by our ordinary perceptual

---

[1] As, for example, in Broad II or Russell IV (Roman numerals refer to works listed in the Bibliography; Arabic numerals to sections in this book).

statements, but this can hardly be right. Philosophical theories of perception tend to be notable for their departure from common sense. They might be interpreted as theories about what our ordinary perceptual statements ought to mean if they are to be true, but in that case the question of the precise sense in which they are true would itself depend on some decision as to which theory of perception was the correct one. Philosophical theories of perception are best thought of not as answers to 'What do we mean when we say we perceive something?', nor even as answers to 'What is it to perceive something?', but rather as answers to 'What sorts of things do we perceive, and how are they related to the sorts of things we ordinarily think of ourselves as perceiving?' This question may not look philosophical, but it becomes philosophical insofar as it is, and can only be, answered not by reference to any empirical facts, but by some form of theoretical argument. Indeed we shall see that the question of a choice between the theories is, in a familiar sense, largely a metaphysical question.

We can say, then, that the analysis of the concept of Perception raises two, related, questions: What is it to perceive something? and What sorts of things do we perceive?[1] Nevertheless the theory that we do not perceive external objects, but only 'ideas' in our minds, is now widely discredited, and with it falls the importance of perception as a topic in the Philosophy of Mind. If, as is widely thought, we are free at last from the confusions and errors of our predecessors, then it might seem that the traditional problems of perception are only of historical or scholastic interest. But this is to forget the other main source of interest in the subject. Epistemology, the Theory of Knowledge, is concerned with the origin, nature, extent and certainty of our knowledge. It seems obvious that it is by perception, and by perception alone, that we are aware of the world around us, so it seems plausible to say that it is by perception, and by perception alone, that we come to know of the existence and nature of that world. So in considering our knowledge of the 'external world' the philosopher will want to consider the way in which perception can and does provide us with that knowledge, and the nature, extent and certainty of the knowledge it provides us with.

In the first part of this book I will be concerned with the prob-

---

[1] There is another, neglected, question, the question of what distinguishes the different senses from one another. For an interesting discussion, cf. Grice II.

INTRODUCTION

lems of the analysis of perception, the questions of what it is to perceive something, and of what sorts of things it is that we perceive. In the second part I will be concerned with epistemological questions concerning perception, in particular the question of how our perception provides us with the knowledge that the things we perceive exist independently of us. These two topics—the analysis of perception and the analysis of empirical knowledge—are related, but not, I think, so closely as has been supposed. I feel that nothing but good can come of the attempt to keep them apart. The traditional approach to perception combines two questions: Do we perceive external objects? and How do we know that we perceive external objects? One question leads to the other, but it would be a mistake to think that an answer to one is an answer to the other. We may have to answer the first before we can answer, or even ask, the second, but that is a different matter.

#### 1.2 SOME CRUCIAL DEFINITIONS

I will have a lot to say about perceiving, external objects, sense data and percepts, and much of what I say will turn on the sense I give these terms, which is not always the sense that philosophers commonly—and in my view misleadingly—give them. So I must begin by explaining what I take 'perceive', 'external object', 'sense datum' and 'percept' to mean. Notice that when I talk of a word having different 'senses' I do not necessarily mean that it is ambiguous, as 'bank' or even 'sense' itself are (the sense of hearing; the sense of 'good'). We might say that 'good' in 'Jones is a good man' has a different sense, depending on whether it is said by the Chaplain or the Captain of Cricket, but this does not mean that 'good' is ambiguous, has more than one meaning. We can draw a rough but ready distinction between 'sense', largely a matter of context, and 'meaning', largely a matter of dictionary definition. And I will want to say, for example, that a man who talks about seeing something in his mind's eye is using 'see' in a different sense from that in which I talk about seeing the door in front of me.[1]

I take *'perceive'* to name the genus of which 'see', 'hear', 'taste', 'smell', and 'feel' name the species. Perhaps it is worth repeating

---

[1] On the other hand I would not want to say that 'see' has different senses in 'I see a twinkling speck', 'I see an enormous star', and 'I see the surface of a star'. On this I agree with White IV.

that the verb which stands to the sense of touch as 'see' stands to vision is not 'touch', but 'feel'. I can touch something without perceiving, feeling, it, and inanimate objects touch one another when it would be absurd to say that they feel one another. Presumably it is because 'feel' has so many different senses, even different meanings, that philosophers sometimes prefer the verb 'touch'. Not only do we talk about feeling pains, which does not involve perception via the tactual sense, but we also talk about feeling sick, dizzy or stupid, which hardly involve perception at all. Moreover, we often use 'feel' not in a sense parallel to 'see' or 'hear', but in senses parallel to 'look at' or 'listen to' ('He felt the fabric carefully'), 'look for' or 'listen for' ('He felt for the switch'), and even 'looks' or 'sounds' ('It feels hard and rough'). But despite all these various uses the fact remains that to say that a person perceives something via the sense of touch, is to say that he feels it.

Now it is widely held that to say that someone perceives something is to say, i.e. it entails, that that thing exists. This seems to me an obvious error. It may be true, in a way, to say that Macbeth didn't perceive a dagger, but this is not true in the way that philosophers take it to be true. They usually take it to mean that Macbeth wasn't perceiving, that he didn't *see* a dagger (or anything else), that he only thought he saw, or merely 'saw' in some special scare-quotes or Pickwickian sense of the verb, a dagger. Obviously Macbeth didn't see *a dagger*, at any rate not a real dagger, but he did see something, something which he described as a dagger. Macbeth did see something, and that something did not exist. The remark that he didn't really see a dagger, that he only 'saw' one, would no doubt gain the acceptance of the man in the street, but that doesn't show that Macbeth didn't see something in the ordinary sense of 'see'. All that it shows is that *if* we draw a distinction between 'see' and 'really see', or between 'see' in a scare-quotes and 'see' in an ordinary sense, and say that Macbeth didn't really see, only 'saw' a dagger, *then* the point of this distinction will be obvious and the usage readily understandable. It does not show that ordinary language makes any such distinction, and does not show that people ordinarily object to saying that Macbeth saw something. They may object once the distinction is drawn, but that is a different matter. So, in the ordinary sense of the word, as opposed to the refined and restricted sense it is given when it is distinguished from a scare-quotes sense of 'see', it is true to say that Macbeth saw, perceived, something which he described as a dagger, even though it would not be true to say that any such dagger existed.

It might be said that the point of the scare-quotes use of 'see' is to enable us to talk about situations where we would ordinarily *deny* that the person saw whatever it was. Only in this way could the philosopher maintain that his scare-quotes term somehow represents ordinary usage. Thus '*A* knows *p*' entails the truth of *p* because if *p* is shown to be false we conclude that *A* didn't know, indeed couldn't have known, *p*. But we might, if we were so moved, introduce a scare-quotes sense of 'know', in which we could talk of a person's 'knowing' something, even though it is false. However the relation between perception and existence is quite different from that between knowledge and truth. If it turns out that the thing in question doesn't exist we don't insist that it wasn't seen, much less that it couldn't have been seen. Macbeth knows, and we know, that he did see something, although in so far as he knows there is no dagger there he knows that he didn't see a dagger, or at any rate not a real dagger (cf. 6.2). What we deny is not that Macbeth *saw* a dagger but that he saw *a dagger*; if we want to put scare-quotes around anything we had better put them around 'a dagger'.

This may seem a trivial verbal point. You may say: if he wants to use 'perceive' in such a sense that it does not follow from the fact that something is perceived that it is really there, then let him, whether that usage be ordinary or technical. But the trouble is that philosophers often use their restricted definition of 'perceive' to escape or side-step important facts. People can and do perceive things which do not exist, and this has important consequences for any theory of knowledge which maintains that it is from what we perceive that we tell whether things really exist. The temptation is to insist that it is logically impossible to perceive things which don't exist, and so side-step the difficulty. Thus, for example, Armstrong[1] and Brown[2] tell us that in hallucination nothing is perceived, meaning by this not that we must use another word to describe what happens, which is all their definition entitles them to say, but that in hallucination we are not in any sense aware of anything. Obviously this 'definitionist sulk'— defining your terms in such a way that you refuse to talk about a

[1] I, p. 83.   [2] I, p. 182.

particular possibility—does nothing to alter the facts of the matter, or solve any problems those facts raise.

However there is a possibility[1] that hallucinations are not, after all, perceived in the way that I perceive the wall in front of me. It may be that hallucinations are nothing but mental images which we mistakenly take to be, and be like, the objects of ordinary veridical perception. Consider dreams and day-dreams: we ordinarily talk as if in dreaming and day-dreaming we were seeing things in the way that I am, at this moment, seeing a desk, a telephone and a typewriter. But a little reflection shows that with day-dreams this is seldom, if ever, the case; seeing a house in a day-dream is more like imagining a house than actually seeing one. And much the same may be true of dreams. When we wake we feel inclined to say that we were seeing lions and tigers just as we are now seeing bed-spread and wallpaper, but the unreality of dreams suggests that what we 'saw' were more like mental images, i.e. that it wasn't a case of seeing things which did not really exist so much as a case of thinking we saw things which we did not really see, things which we 'saw' only in the rather attenuated sense in which one 'sees' images. It is just a short step from this to the suggestion that what happens in hallucinations, DTs, etc., is not that we *see* things that are not there, but that we *think* we are seeing things when in fact we are only imagining them, are 'seeing' them only in the way that we 'see' images in memory or imagination. If this should turn out to be the case then we must agree that Macbeth didn't see, in the ordinary straight-forward sense of 'see', anything at all; we would have to agree that Macbeth merely thought he saw something. But the question of whether or not this is the case is an empirical question, and is not to be settled *a priori*, by a mere choice of definitions. If it is the case that we do not see hallucinations in anything like the way I now see a desk, a telephone, and a typewriter, then I am mistaken in insisting that Macbeth saw, in the ordinary sense of 'see', something which he described as a dagger—and so too is Macbeth himself, Shakespeare, Sir Francis Bacon, and the ordinary man. So for simplicity I will assume that this suggestion about hallucinations is incorrect, for it seems to me that our ordinary use of language assumes that it is incorrect, i.e. it seems to me that we ordinarily think, rightly or wrongly, that a man perceives hallucinations in just the same way

[1] Cf. Britton I, Hirst I, pp. 41–5 ff.

INTRODUCTION                                                          19

he perceives real physical objects. And whatever the truth is about hallucinations there are some things which we do perceive, in the straight-forward sense of 'perceive', which do not really exist, e.g. after-images.

As well as the fact that in the large majority of cases what we perceive does really exist, so that '$A$ perceives $X$' normally implies '$X$ really exists', there are two mistakes which explain the tendency to take it for granted that '$A$ perceives $X$' entails that $X$ exists. The first is the confusion between denying that Macbeth saw a dagger and denying that he saw anything at all. This confusion is compounded by the fact that '$A$ thought he saw $X$' can be interpreted in two different ways. It might mean that $A$ did see something which he mistakenly thought was $B$, as when, alone in the jungle at twilight, I see a vine and think it is a snake. Or it might mean that $A$ thought he saw $B$ when in fact he saw nothing at all, as when a child thinks he sees a bear in a pitch-dark room, or a lunatic thinks he sees the French army advancing across the fields of Waterloo behind him. Only in the second case does '$A$ thought he saw $X$' mean that $A$ didn't see anything at all. But, given that the above suggestion about hallucinations is incorrect, 'Macbeth thought he saw a dagger' would be false in both interpretations. He did see something, which he described as a dagger, but it is very doubtful whether he really thought that something was a dagger. The Shakespearean evidence seems to be that he knew it wasn't.

The second mistake involves a confusion over the meaning of 'exist'. When we are talking about what we perceive there are two types or kinds of existence we might have in mind. When we say of a table or a chair or a sound that it exists, that it is really or actually there, we mean, among other things, that it exists, is there, whether we or anyone else perceive it or not. But when we say of an after-image or a pain that it exists we do not mean that it exists, is there, whether it is perceived or not, but only that it is perceived. I will distinguish between these two by talking about *real existence* and *perceived existence*. To say that something really exists is to say that it exists independently of our perception of it; after-images do not really exist, in this sense. To say that something has perceived existence is to say that it exists inasmuch as it is perceived; after-images exist in this sense and only in this sense. The difficulty is that we talk about tables as existing and after-images as existing as if we were saying the same thing about each. Naturally it follows from the fact that something is perceived that it exists in some sense, but in this sense even hallucinations exist. Oddly enough philosophers who refuse to speak of Macbeth seeing anything, on the grounds that what he 'saw' did not exist, have usually been quite prepared to say that we see after-images, even though Macbeth's dagger existed every bit as much as after-images do!

A thing which really exists, in my sense, I call an *external object*. Since I am concerned only with what we perceive, I will restrict this term to things which can be perceived by the unaided ear, eye, nose, etc., thus eliminating atoms, molecules and the like. Physical objects, as ordinarily thought of, are external objects; so too are sounds, smells, shadows, clouds, puddles and perhaps even rainbows. I think this definition of 'external object' follows accepted philosophical usage, except perhaps for the fact that it includes my body, and thus myself, as an external object. External objects are, by definition, non-sense-dependent. They might also be said to be 'physical', in the sense that they can be assigned a spatial location, even if only an indeterminate one of the kind we assign to sounds or smells. And they might also be said to be 'public', in the special sense that they can be perceived by more than one person. It seems to me that the three distinctions between external and sense-dependent objects, physical and mental objects, and public and private objects, are three different distinctions, although it may be that certain of these characteristics entail certain others.

In philosophical theories of perception external objects are contrasted with sense-dependent entities known variously as ideas, sensations, sense-impressions, sense data, etc. I will use the term '*percept*'. A percept is something which exists only in so far as it is perceived, like an after-image or an hallucination or possibly a mirage. The main feature of certain theories of perception is that they claim that we always and can only perceive percepts, i.e. that we never perceive external objects, except perhaps in the extended sense that we perceive percepts 'of' external objects. These sense-dependent percepts might also be said to be mental, in the sense that they have no spatial location, and private, in the special sense that they can be perceived by only one person, the person whose percept it is.

INTRODUCTION 21

There has been a tendency to use 'sense datum' as the name for what I call percepts. This has been a source of great confusion. For whoever may have invented the term[1], those who have done most to make it part of the philosphers' vocabulary have always insisted that it is a *theory-neutral* term, that talk about sense data does not commit us to any particular theory of perception. Thus Price[2] 'It is meant to be a *neutral* term. The use of it does not imply the acceptance of any particular theory'. Broad[3] says that is is held 'to be part of the *meaning* of a sense datum to be private and mind-dependent. Now this is certainly no part of what I mean by the word, and it is obviously no part of what Mr Russell means by it'. Like Broad Moore always allowed it possible for sense data to exist unperceived. In the earliest work in which he uses the term he was even prepared to allow they might exist unperceived in the mind,[4] and although by the time of his last published thoughts on the subject[5] he had decided that visual sense data are not identical with parts of the surfaces of physical objects it is obvious that he still regarded this as a factual matter, and not a matter of definition. Moore's typical procedure was to try to explain, by some more or less ostensive method, what he meant by a sense datum and then to ask such questions as 'Do they exist when not perceived?' or 'Are they or do they include parts of the surfaces of external objects?' It would be silly to ask such questions of percepts, for the definition of 'percept' is such that the answer must be negative. What Moore is asking, in effect, is whether sense data are percepts, whether the sense data whose existence he correctly took to be undeniable are special entities which exist only when perceived and which, in a sense, come between us and the things we ordinarily think of ourselves as perceiving, or whether they are or include parts or aspects of external objects. When Paul wondered whether there was a problem about sense data[6] what he was suggesting, in effect, was that we mustn't let the possibility of describing what we perceive in certain special, restricted, ways mislead us into thinking that what we describe in these ways are entities of a special sort to be distinguished from the cars and cats, windows and walls, we would ordinarily describe ourselves as perceiving.

If this were a historical work it would be necessary to add a

[1] Cf. Hall I.   [2] I, p. 19.   [3] I, pp. 211–12.
[4] IV, p. 54, cf. p. 31.   [5] V.   [6] I.

chapter on the Decline and Fall of the Sense Datum. It seems that Moore spoke so much of sense data because he hoped that an examination of them would enable us to decide between Realist and non-Realist, or 'percept', theories of perception. Price said that 'sense datum' is a theory-neutral term, but nevertheless he seems to think it logically impossible for the redness I see when I see datum to be the redness of the tomato, the external object. The existence of the redness is, he thinks, 'as certain as anything can be', whereas the existence of the tomato is doubtful, and 'how can a certainly real quality qualify a doubtful real entity? Plainly it cannot'.[1] Finally Ayer makes no bones about it: 'If we accept the sense datum terminology, then we must reject the terminology of naive realism for the two are mutually incompatible'.[2] It is unfortunate that those since Moore who have most wanted to speak about sense data have been their own worst enemies.

So we find contemporary philosophers rejecting the notion of a sense datum when their real target is not sense data but percepts, Locke's ideas, Berkeley's sensations. A valuable philosophical term —although its value lies in epistemology rather than in discussing the theories of perception—is in danger of being lost through confusion if we do not carefully insist upon the distinction between sense data, which are the 'immediate' objects of perception and so exist whenever we perceive, and percepts, which are entities which exist only in so far as they are perceived. I will leave the full explanation of sense data until Chapter 11, but for the moment we can notice two points. First, sense data are, so to speak, the theory-neutral equivalent of percepts; they are what the Idealist and the Causal Theorist will identify as percepts. The question at issue between the theories of perception is, in part, whether sense data are percepts, whether the things we are fundamentally aware of in our perception are things which exist only in so far as we perceive them. And second, if Realism is correct than talk about sense data will, for the most part, simply be a rather special way of talking about external objects as we happen to perceive them. If Realism is correct sense data will not form some separate class of special items which have to be distinguished from tables and chairs, sounds and smells. But even if there is no such class, the term 'sense datum' may still have a point and a meaning, in just the way that 'curiosity' (in the sense of something we find curious)

[1] I, p. 105. I come back to this argument on p. 39 n 3 below. [2] II, p. 48.

INTRODUCTION 23

has a point and a meaning, even though there is no separate and special class of curiosities which have to be distinguished from tables and chairs, sounds and smells, and the other things we find curious. We must beware of the *unum nomen—unum nominatum* fallacy.

# PART ONE
# PERCEPTION

CHAPTER 2
# AN ANALYSIS OF PERCEPTION

## 2.1 SENSORY AWARENESS

What is it to perceive, what is involved in perceiving, something? In so far as this is a philosophical question, our task is to distinguish, conceptually, the different elements involved in perceiving things. We might also distinguish perception (seeing, hearing, etc.) from such related things as observing (looking at, listening to, etc.) and examining or searching (looking for, listening for, etc.).[1] I shall not be concerned with these questions. They may be interesting in their own right, but I do not see that they throw any light on the traditional problems of perception, as is sometimes suggested.

The first and most important element in perception is the basic process, activity, state, of awareness, by which we are acquainted with the items we perceive. This activity, state, process—I am not sure what to call it—is sometimes referred to as 'sensation', as opposed to 'perception' in the full sense. This can be very misleading (cf. 5.2), so I prefer to speak of 'sensory awareness', using the verb 'to sense'. It is difficult, perhaps impossible, to provide a verbal analysis of this sensory awareness. It is that feature common to all cases of perception, the element of being aware of, acquainted with, items in the special way we are when we perceive them. The difficulty of explaining this in words is very like the difficulty of explaining in words what a colour, e.g. red, is. Perhaps the best we can do is provide an empirical account of what sensory awareness is, saying, for example, that sensory awareness is what occurs when the brain is activated in certain ways, just as red is what is seen when the retina is stimulated by light waves of a certain sort. This is, of course, not an analytic definition of 'sensory awareness', for it is a contingent fact that sensory awareness is connected with certain types of brain activity, just as it is a contingent fact that the perception of red is connected with the retina's being stimulated by light waves of a certain sort. But I don't see how there can be an analytic definition of 'sensory awareness' which does not make use of terms like 'sensing',

[1] For a discussion cf. Sibley I.

'acquaintance', 'awareness'—terms which are themselves equally in need of definition.

It seems obvious to me that perception involves some such process of awareness, and that any attempt to analyse perception without it must be grossly implausible. Even so there have been two recent attempts. The first is that of Ryle[1] who argues that perception is not a process at all. Rather it is an 'achievement', something which takes no time. In the next section I shall argue that it is noticing, rather than perception as such, which is the achievement, but I think we can also see that Ryle's own account itself presupposes the existence of a process of some sort. His suggestion is that it is not perceiving, but observing, which takes time. He does not explain what he means by 'observation'— instead he concentrates on what it is to recognize, e.g. a tune or a thimble—but the natural thing to say is that observing involves paying attention to what we perceive, that the difference between perceiving $x$ and observing $x$ is that in the latter case we attend to what we perceive. Now Ryle also wants to argue that attending is not a process either. Attending is not itself something that we do, rather it is a way of doing something else. But quite apart from any difficulties about what could be meant by a way of doing something that takes no time, there is the objection that, on this account, neither attending nor perceiving are processes, take time, while observing, which seems to be perceiving plus attending, does take time!

The second attempt to analyse perception without reference to a process of awareness is that of Armstrong.[2] He argues that the concept of perception is 'a complex concept, definable in terms of such concepts as knowledge, belief and inclination to believe',[3] 'perception . . . is the acquiring knowledge of, or inclination to believe in, particular facts about the physical world, by means of the senses, normally accompanied by knowledge of the means'.[4] Apart from difficulties of detail in Armstrong's theory, there seem to me to be four main objections.

First, his analysis seems circular. Perception is *nothing but* the acquiring of knowledge of particular facts about the world by means of the senses'.[5] But surely the way in which we acquire knowledge by means of the senses is by perceiving? Armstrong seems to be saying that perception is nothing but the acquiring of

[1] II, ch. 7; cf. I, ch. 7.　[2] I.　[3] I, p. 121.　[4] I, p. 114.　[5] I, p. 112.

knowledge of particular facts about the world by perceiving things in that world![1]

Second, this analysis means that it is logically impossible to perceive something without forming, or being inclined to form, any beliefs about it. But surely one can, for example, ascribe perception to animals without committing oneself as to whether they have beliefs or not?

Third, this analysis makes it impossible to explain how we acquire knowledge and belief about the physical world. Armstrong says that we need not give an explanation for everything; we must not be like children who go on asking 'Why?'.[2] But this answer will not satisfy anyone who is interested in the theory of knowledge, and the plain fact of the matter is that it is not we who are being silly in demanding an answer to questions like 'How do you know there is a book there?', but Armstrong who is being silly in refusing to allow the obvious answer 'Because I see it'. On Armstrong's account to say 'I know there is a book there because I see it' is equivalent to saying 'I know there is a book there because I have acquired the knowledge that there is a book there'![3]

[1] For fuller discussion of this point cf. Nelson I.  [2] Cf. I, pp. 93, 120, 133.

[3] Armstrong thinks that 'Because I see it' is not an answer to the question because 'in order to be justified in passing from the perceptual experience to a certain belief about physical reality, we should have to know that certain perceptual experiences generally occurred when a certain state of affairs obtained in the physical world. But to know this we would have to gain independent knowledge of the physical world; and if our perceptual experience is the basis of our knowledge of the world this cannot be done' (p. 116). There is, of course, an epistemological problem that we will have to consider, the problem of whether and how perception can provide us with the knowledge that our perception is, on a particular occasion, veridical. But it seems foolhardy rather than courageous to try to avoid the problem by denying that perception does provide us with knowledge about the world, by saying that perception is not the source of such knowledge but simply the, apparently inexplicable, acquiring of such knowledge. As for Armstrong's difficulty: (1) It is not true that in order to know that my perception is veridical (that when certain perceptual experiences occur a certain state of affairs obtains in the physical world) we have to have some independent knowledge of the physical world. It should be a familiar fact that I discover that what I perceive is really there not by some independent, non-perceptual, means, but by noticing what I and other people perceive by this and other senses at this and other times. (2) It is not at all clear how Armstrong avoids his own difficulty, how he explains my knowledge that my perception is veridical. I take it that he does not want to say that we do have some independent means of knowing this, nor, I hope, that this is known by intuition. But if he is prepared to say that our perceptual experiences can justify us in passing from perceptual experiences to a certain belief about physical reality, what becomes of his original argument?

Fourth, the analysis commits Armstrong to the coherence theory of truth which he himself rejects. On his theory veridical perception consists in the acquiring of knowledge about a certain thing or things, those that we are said to perceive. But how are we to tell whether this is knowledge or merely a mistaken belief, if we can never examine the things in question to see whether what we believe about them is in fact true? All that such 'examining' can be, for Armstrong, is the acquiring of further beliefs or knowledge about them, since that is all perception is. The sole test of truth, then, seems to be whether one set of beliefs fits in with another. We can acquire further beliefs and see how they cohere with our original ones, but we can never test these beliefs against the things themselves.

## 2.2 NOTICING

It is often suggested that perception involves not just a sensory element of awareness but also an intellectual or cognitive element of judging or something of that sort. Now perhaps I cannot be said to perceive, in the full sense, a tree unless I realize, recognize, 'judge', that it is a tree, but clearly there is a sense in which I perceive the tree even if I do not realize that that is what it is. If I see a vague shape in the fog which in fact is a tree, I can be said to see the tree even if I do not know what it is, even if I mistakenly take it to be a man. What is necessary if I am to be said, in any sense, to see the tree is not that I 'judge' it to be a tree, but that I *notice* it. The point is that I may notice the tree without noticing *that* it is a tree; I will come back to 'noticing that' in the next section.

Sometimes the question of whether I noticed something is the question of whether I took special note of it, whether I was struck by it or paid it special attention. But at others it is merely the question of whether I happened to recognize it, or realized that it (whatever it was) was there. I am here using 'notice' in the weakest possible sense, and noticing, in this sense, is a necessary element in perception. If I do not notice the stain on the carpet, in the sense that I do not even realize that it is there, then I cannot, in any sense, be said to perceive the stain on the carpet. Indeed, if the above account of sensory awareness is correct, i.e. if sensory awareness occurs whenever the brain and sense organs are activated in their various ways, then noticing is what turns sensory awareness into perception. Suppose I glance at the cover of a book without noticing the name of the publisher written across the bottom. In so far as my physiological and neurological state is affected not just by the book but by this writing as well—in so far as my retina is stimulated by lightwaves reflected by this writing—then we should, on the account given, say that I have sensory awareness of the publisher's name. But since I do not notice the name I do not perceive it. In the same way we might say that in subliminal perception, or in psychic (or post-hypnotic) blindness, we have sensory awareness without noticing, and hence without perceiving. To perceive an item, in any sense which means that I am conscious of it, I have not only to sense it, I must also notice it.

Noticing is what Ryle would call an achievement, what White[1] calls a reception. That is, it is not something that takes time. We may take time *to* notice something just as we take time to score a goal, but we do not spend time *in* noticing any more than we spend time in scoring the goal.[2] There is but one possible exception to this general rule: where what is noticed itself takes time. If I say 'I noticed him move slowly across the room, pick up a paper, glance at it, open the door, and go out', it is a moot point whether we say that noticing this lasted as long as it took to happen, or say that the noticing itself took no time but occurred at the precise moment when the events concluded. However, it is because noticing is an essential element in perception that the perceptual verbs 'see', 'hear', etc., are characteristically used as 'achievement' verbs. As Ryle points out, we can say 'I have seen it' as soon as we can say 'I see it', just as we can say 'I have won it' as soon as we can say 'I win it'. Moreover the perceptual verbs are seldom used in continuous tenses or with temporal expressions. This is because 'see' and the other verbs can usually be replaced by something like 'notice visually'.

Nevertheless these verbs are not always used in an 'achievement' sense. As Sibley points out[3] we can always ask 'Did you see it for long?', when this question should be out of place if 'see' is an 'achievement' verb. What this shows, I think, is not, as Sibley suggests, that 'see' never has an 'achievement' sense, but that we can always pass from an 'achievement' sense to an 'non-achievement' sense, where we refer not so much to the 'achievement' of

[1] III, ch. 3.   [2] Cf. Ryle II, ch. 7.   [3] I, p. 472.

noticing as to the process of sensory awareness. In other words noticing always requires a process of sensory awareness, and verbs like 'see' can refer to the latter as much as the former. They are, on occasions, used with temporal expressions and in continuous tenses: I am seeing stars; Are you sure you are seeing what I am seeing?; Do you still see it?; I saw it for a few moments and then it went behind some clouds; etc. And although it is true that we can say 'I have seen it' as soon as we can say 'I see it', it is also true that we can say 'I have seen it, and still see it' in a way we cannot say 'I have won it, and still win it'. 'I have seen it' refers to the noticing; 'I still see it' or 'I am now seeing it' to the sensory awareness.

## 2.3 PERCEPTION-THAT

I will call the third element in perception 'perception-that'. I shall even be tempted to speak of 'misperception-that', although I will try to avoid this barbarism by talking instead of what a person takes or judges himself to perceive. To say that a person judges, in this sense, that what he perceives is, say, a flower, is not to say that he forms an explicit judgement, says out loud or to himself, that this is a flower. Rather it is to say that he takes what he perceives to be a flower. One can think, even know, that what one perceives is a flower, without thinking 'It is a flower'.

Perceiving-that differs from noticing in that it involves knowing something, or at least having some opinion, about what is perceived. To notice a flower is not necessarily to know or think anything about what it is or is like, although it may be that one cannot notice something without at the same time noticing something about it, perceiving that it is of such and such a kind. We should also distinguish between what a person takes what he perceives to be, and what he perceives it to be. If, because of the unusual lighting, something which is white looks vermilion I sense and perceive it to be vermilion, but I may not take it to be vermilion, perceive that it is vermilion, if only because I do not recognize the colour.

What precisely is it to perceive that? It is tempting to say that I perceive that this is, e.g. a scarecrow if and only if I know or realize or recognize that it is a scarecrow. But this will not do for I may very well know that it is a scarecrow because someone has told me or because I have seen it before, without our wanting to say that on this occasion, when all I see is a vague looming shape, I perceive that it is a scarecrow. Instead we might say that to perceive that $p$ is to know that $p$ because of what one is, at that time, perceiving, which means I do not, on this occasion, perceive that it is a scarecrow because although I know that it is a scarecrow I do not know it because of what I now perceive. But even this will not do, for two reasons: When I perceive the scarecrow in broad daylight we would still want to say that I perceive that it is a scarecrow, even if I already know that it is. And a person can make a mistake of perception-that, can take or judge what he perceives to be something that is not, in which case, of course, he cannot be said to know that $p$. So let us say that to perceive that $p$ is to consider $p$ to be true of the things I notice, where either I consider $p$ to be true because I now notice them, or I would consider $p$ to be true for this reason if I did not already consider it true for other reasons.

By means of this notion of perception-that we might distinguish a strong and a weak sense of 'perceive'. When I see the vague looming shape in the fog there is a sense in which I do not see a scarecrow, but only a vague looming shape, and a sense in which I do see a scarecrow, since that is what the vague looming shape is. In a strong sense I do not perceive the scarecrow because I do not perceive that it is a scarecrow, but in a weak sense I do perceive a scarecrow. Perception-that is necessary for perception in the strong sense, but not for perception in the weak sense.

We might also notice that some perception-that goes beyond perception. If I see your hat in the hall I may say that I see that you are in, but this perception-that is not an element in my perception because I do not perceive you. Perception only includes such perception-that as does not go beyond what is perceived. When I 'judge' that you are in I am making a 'judgement' not about your hat, which I do notice, but about you, whom I do not notice, and so this 'judgement', this perception-that, cannot be described as an element in my perception.

It is this element of judging, realizing, perceiving-that which philosophers have referred to when they insisted that all perception involves judgement. It also explains how what we perceive can, in a sense, depend upon our training and experience. When faced with the word 'Edinborough' the adult, the child and the Eskimo see, respectively, a misprint, the name of a city and a meaningless collection of marks, even though, in a definite sense, they all see the same thing.[1] The differences lie in their perception-that, in what they notice *about* what they notice, and not in what they perceive. They sense, notice and perceive the same things, but each, so to speak, makes something different of it.

To what extent does perception-that involve knowledge? We might insist that we cannot be said to perceive that $p$ unless $p$ is true, but this 'definitional stop' would not affect the fact that we often make mistakes about what we perceive, and we need some way of describing this situation. To take yourself to perceive $x$ is not to know that you are perceiving $x$, for you might be mistaken. Indeed even if what is perceived is $x$, as we take it to be, it doesn't follow that we know that it is $x$. Warnock[2] argues that we cannot be said to see that $p$ unless we know that $p$. But suppose I am in a fairground 'house of illusion' where nothing, or very little, is what it seems, so that we insist that just looking is never sufficient to give us knowledge. I see what I take to be a tomato, and although it might be a billiard ball it really is a tomato. Here I see that it is a tomato, and it is a tomato, but I do not know that it is a tomato. So perception-that does not necessarily involve knowledge of what is perceived, although obviously, and this is important, it involves knowledge of what we take what we perceive to be.

[1] Cf. Ryle III, p. 436.         [2] II, cf. I, p. 139.

# CHAPTER 3
# THE PHILOSOPHICAL THEORIES OF PERCEPTION

## 3.1 THE THREE THEORIES

Our next question is: What is the nature and status of the things we perceive? The question is asked with the distinction between percepts and external objects in mind, and the expected answer is either that we sometimes perceive external objects or that we always perceive percepts. It would clearly be wrong to say that we always perceive external objects. People do suffer hallucinations and do perceive after-images, and these are paradigm examples of percepts.

If it is held that we always perceive percepts it is possible to add that there exist external objects as well as percepts, even though we never perceive them except in the sense that we perceive percepts caused by them, or to add that there are no external objects. In this way we get the three traditional theories of perception: the Realist theory that we can and usually do perceive external objects; the Causal theory that we never perceive external objects, but only the percepts they produce in our minds; and the Idealist theory that we perceive nothing but percepts and there are no external objects at all. These are not the only possible theories—Causal theories may be sub-divided into Representative and non-Representative theories, depending on whether it is held that our percepts resemble or represent external objects or not, and there is also the Sensibilia theory that we will discuss in 9.2—but they are the most common and the most plausible, and it is with them that we shall principally be concerned.

The difference between the three theories can also be brought out by considering what they have to say about physical objects, which as we saw (1.2) are to be distinguished from external objects. It would be grossly implausible to say that we do not perceive physical objects, such as tables and chairs, so philosophers have usually preferred to say that we do perceive them, while giving this phrase a special interpretation. The Causal theorist gives a special interpretation to 'perceive'; according to him we perceive physical objects only in the sense that we perceive their effects, the percepts they produce in our minds. The Idealist gives a special interpretation to 'physical object'; according to him physical objects are not external objects, do not exist independently of our perception of them, and to perceive a physical object is just to perceive the appropriate sort of percept.

These, then, are the three major philosophical theories of perception (Phenomenalism, as we will see in the next chapter, is not, strictly speaking, a theory of perception at all). The usual procedure of philosophers discussing these theories has been either to reject Realism by means of arguments designed to show that we perceive nothing but percepts, or to reinstate Realism by showing that these arguments are fallacious or inconclusive. And since contemporary philosophers tend to favour the second alternative their discussions of perception often seem entirely negative, not to say inbred, concerned solely with the mistakes of others. Locke and Berkeley may have been wrong, but at least they had something novel and constructive to say; contemporary discussions seem to have nothing original or positive to offer. I think the moral is that if Idealism and the Causal theory are to be rejected, then the traditional controversies become of merely historical interest, and our attention should turn instead to the epistemological questions about perception. These questions have usually been asked with reference to the traditional theories of perception, but now they should be separated from them. Even a purely destructive criticism of those theories will be of positive value if it helps clarify the epistemological issues.

In the chapters that follow I shall consider the various arguments for and against the three theories. I shall argue that none succeeds in proving any one theory as against the others, and that therefore the choice between the theories will have to be made on theoretical or methodological grounds. Four considerations will be particularly important: the extent to which the theories conform to common sense; the extent to which they follow Occam's razor, and restrict the number of entities that need to be postulated; the extent to which they raise or solve epistemological problems; and the extent to which they fit with or are suggested by the various facts about perception. It is obvious from the outset that Realism is preferable from the point of view of common sense, for it is the common sense theory. Equally obviously, Idealism is preferable from the point of view of Occam's razor, since it alone avoids postulating both percepts and external objects. And, as Berkeley showed once and for all, the Causal theory is at a considerable disadvantage from the epistemological point of view. It is usually suggested that Idealism is preferable to Realism on this point also, but I shall argue that this is not so. Finally, the Causal theory is often thought to be proved, or at least supported, by the physiological and neurological facts about perception; and the facts of 'illusion' and 'appearance' are often cited in support of Idealism.

### 3.2 THE INTRODUCTION OF PERCEPTS

There has always been a tendency to take it for granted, as obvious and undeniable, that whenever we perceive we perceive percepts. Thus Berkeley, apparently driven by the problem of how we can know about what we do not perceive, takes it as obvious that things exist only in so far as they are perceived,[1] although the reception of the *Principles of Human Knowledge* seems to have shown him that this is not so obvious after all, and that the more detailed argument of the *Three Dialogues* is necessary. Again, some philosophers may have been worried by the rather strange problem of how the mind can be acquainted with what is not mental, and so decided that it must be acquainted with things that are mental, i.e. percepts. Or they may have slipped from the fact that perception itself is, in some sense, private, mental and sense-dependent, to the conclusion that what is perceived must also be private, mental and sense dependent.

Another important factor here is what has been called 'the reification of appearances'. This connects with what I will call the argument from appearance (6.5), the argument that since we cannot perceive anything without perceiving the appearance it presents to us, it follows that we can never perceive the things themselves, but have to be content with their appearances. This is to draw an untenable distinction between things and their appearances and, in effect, to treat appearances as if they were objects in their own right. An appearance is not something over and above, or separate and distinct from, that of which it is the appearance. If I compliment a girl on her appearance I do not expect her to be insulted because I have complimented her appearance and not her. To treat appearances as objects would seem to be a paradigm

---

[1] I, § 6.

example of a category mistake, and, so far from its being the case that we can never perceive things but only their appearances, the truth is that if we cannot perceive things we cannot perceive their appearances either, since to perceive a thing's appearance is itself to perceive that thing.[1]

But perhaps the most important source of error here is the confusion between sense data and percepts. If we do not realize that 'sense datum' is a theory-neutral term, that the acceptance of the (analytic) truth that all perception involves perception of sense data does not commit us to any particular theory of perception, then it will be easy to construe Moore's attempts at a more or less ostensive explanation of sense data as attempts to show, ostensively, that all perception is perception of percepts. This was not Moore's intention. He tried to introduce sense data as a preliminary to asking, not as a way of answering, such questions as 'Does what I perceive continue to exist when I close my eyes?'. Saying that all perception involves perception of sense data is not saying that we never perceive anything but sense data, any more than saying that all jumping involves jumping jumps is saying that we never jump anything but jumps, that we cannot jump fences or streams. Surprisingly, it is even possible to insist that sense data exist only when perceived, and still adopt a Realist theory of perception (cf. 11.4).

This point, that the introduction of sense data is not the discovery of a new sort of entity but merely the introduction of a new way of talking about what we perceive, was clearly made by Paul.[2] He points out[3] that this way of talking, this sense datum terminology, is not necessary; the question is only whether it is useful in helping to solve some philosophical problems about perception.[4] And we must be particularly careful that we do not introduce sense data in such a way that it follows from the definition of 'sense datum' that sense data are percepts. To do this is, in effect, to define percepts into existence. This, if anything, is the 'sense datum fallacy'.

For example: in talking about appearances we may think it convenient to introduce a terminology in which we do not have to keep using such verbs as 'looks', 'sounds', 'appears', etc. So we

---

[1] The latest to reify appearances is Grice I. He finds no oddity or absurdity in suggesting that an object can be the cause of its look, cf. White's reply, II.
[2] I.    [3] I, p. 109, and footnote.    [4] I, p. 108.

decide to introduce a term, 'sense datum', such that when, for example, a white object *looks* red to us we say that the sense datum *is* red. But this suggests that there is something, which we see, which is red. This red something cannot be the object; *ex hypothesi*, the object is white. So we conclude that there is a sense-dependent entity, a percept, which, so to speak, comes between us and the white object we originally said we saw. This terminological trick even led Moore to think that it is 'quite plain that what is meant by saying that the same surface 'looks' different to two different people is that each is ('directly') seeing an entity which *is* different from what the other is seeing'.[1]

Similarly Price[2] decides to use the term 'sense datum' as a name for that in what we perceive which we cannot be mistaken about. But 'How can a certainly real quality qualify a doubtfully real entity? Plainly it cannot'.[3] So Price comes to think of 'sense datum' as the name of a special entity, to be distinguished from the objects we would ordinarily describe ourselves as perceiving, objects about whose existence we are not absolutely certain. But all the definition entitles him to do is to think of 'sense datum' as a term for referring to certain features of what we perceive, be it external object or percept, viz. those features which we cannot be mistaken about.

To repeat: 'sense datum' is a theory-neutral term, and to insist that whenever we perceive we perceive sense data, is not to commit oneself to any particular theory of perception. In particular it is not to say that whenever we perceive we perceive percepts. The question at issue between the different theories of perception is, in part, the question of whether sense data are percepts, or whether to talk about sense data is to talk, in a special way, about the various external objects which we happen to perceive.[4]

[1] V, p. 208.   [2] I, p. 3.
[3] I, p. 105. The argument is invalid anyhow. Presumably to ask whether a certainly real quality can qualify a doubtful real entity is to ask whether something about whose existence I am certain can belong to something about whose existence I am uncertain. The answer is that it can. I am certain that I hear a noise without being certain whether the noise is made by a train, but that does not mean that the noise cannot be made by a train.
[4] Ayer has argued that sense data cannot be said to include parts or aspects of external objects. The arguments are mistaken. At one time (III, pp. 83-4) he held that sense data cannot be identical with physical entities because we can perceive physical entities which do not exist (e.g. Macbeth's dagger), when it is self-contradictory to say of a sense datum that it is perceived but does not exist. But in the sense in which sense data must exist if they are perceived (viz.

### 3.3 THE THEORIES AS ALTERNATIVE LANGUAGES

Paul's explanation of how talk about sense data is, if Realism is correct, simply an alternative, and perhaps philosophically expedient, way of talking about external objects, in so far as we perceive them, might well be called an 'alternative language' account of sense data. This should not be confused, as it is by Hirst[1] and perhaps by Ayer himself, with the very different 'alternative language' account of the theories of perception, as suggested by Ayer.[2] It seems to me that Paul's article would have shown, once and for all, the precise status of sense datum talk, had not matters been confused by Ayer's appearing to take over Paul's position when in fact he was advocating a theory in direct opposition to it.

I suggested earlier that there might be no way of proving one theory of perception as against the others. Certainly it seems they

---

'perceived existence') it is equally true that a physical entity, even Macbeth's dagger, must exist if it is perceived. In the relevent sense of 'exists', it is a contradiction to say that a physical entity is perceived but does not exist. (In a footnote Ayer tries to avoid this objection, as raised by Moore, but the discussion is bedevilled not only by the confusion over 'exists', but also by the confusing use of 'physical entity', which Moore takes to mean 'external object' and Ayer does not.)

More recently Ayer has restated the argument. He says (VI, p. 97) that it is self-contradictory to say of a sense datum that it exists without being perceived, but not self-contradictory to say this of the surface of a physical object, which means that no sense datum can be identical with part of the surface of a physical object. Like the earlier argument this depends on the assumption that if something can significantly be said of $x$ but not of $y$, then it follows that $x$ and $y$ cannot be identical (not even in the sense in which a sense datum might be said to be identical with the surface of a physical object). The assumption is false. Suppose that the man in the corner is the man who broke the bank at Monte Carlo. I can say, significantly but not truly, that the man who broke the bank at Monte Carlo is not standing in the corner. I cannot significantly say that the man standing in the corner is not standing in the corner. But this does not show that the man standing in the corner cannot be the man who broke the bank at Monte Carlo.

By Ayer's argument we could claim that two-dimensional surfaces cannot, in any sense, be parts of three-dimensional objects, since there are things that we can say about three-dimensional objects which we cannot significantly say about surfaces, and vice versa. The point is that we cannot describe something as really existing and still be talking about sense data, just as we cannot talk about something as extended in three-dimensions and still be talking about surfaces. This does not mean that sense data cannot, in some sense, include parts of external objects, or that three-dimensional objects cannot, in some sense, include two-dimensional surfaces. What Ayer's arguments show is that Sense Datum and Physical Object are different *concepts*, and this is indisputable.

[1] I, Ch. 5, § 1.   [2] II, Ch. 1, § 5.

cannot be proved or disproved simply by appealing to the empirical facts, unless one assumes, as Realists naturally tend to, that one's theory is included among the facts. Ayer asks 'Can we discover any empirical evidence that favours any one of these theories rather than another?'[1] and his answer is negative. So it is not surprising that the author of *Language, Truth and Logic* should refuse to regard these theories as genuine theories at all. As Ayer puts it in discussing one particular dispute:[2]

'So long as we persist in regarding the issue as one concerning a matter of fact it is impossible for us to refute him. We cannot refute him, because, as far as the facts are concerned, there is really no dispute between us. . . . Our disagreement . . . consists in the fact that he refuses to describe the phenomena in the way in which we describe them. . . . In other words, we are not disputing about the validity of two conflicting sets of hypotheses, but about the choice of two different languages'.

Because the facts do not and cannot support or prove any one theory against any other theory the difference between the theories is not one of fact. Indeed there is no real conflict at all, merely different ways of describing and reporting the same facts and the only question is: which way is the most convenient?

Now in one sense these theories do not dispute the facts, but in another they do. They do not dispute the facts in that no fact can be discovered which cannot be fitted in with each of the theories; no matter what empirical evidence we produce the Realist, the Idealist and the Causal theories will all be able to accept, account for and describe it in terms of their theories. The theories are not up for empirical confirmation or refutation and it is this which leads a positivist to deny that they are genuine theories at all. But the theorists do dispute the facts in the sense that each insists that something is true which the others regard as false. If $A$ says 'We perceive external objects', and $B$ says 'We never perceive external objects but only their effects', and $C$ says 'There are no external objects to perceive' they are, in an obvious sense, disputing the facts of the matter, i.e. arguing over what is true and what false, even though there may be no empirical way of deciding between them. It is for this reason that Hirst insists[3] that Ayer's

[1] II, p. 53.   [2] II, p. 17-8.   [3] I, p. 116.

'thesis can only be maintained as the result of a peculiar use of the words "fact" and "description" ', although we might agree with Ayer that a fact that cannot be shown to be a fact is a very strange fact, indeed. Perhaps the least confusing way of describing the situation is to say that the theories do not dispute the *empirical* facts but rather the *interpretation* of those facts.

We can now distinguish Ayer's position from Paul's. If Paul were to talk of 'alternative languages' he would mean different ways of saying the same things; for him talk about sense data does not entail the denial of anything we would ordinarily say about what we perceive, it is merely a different, perhaps preferable, way of stating the same things. But Ayer's 'alternative languages' are not equivalent ways of saying the same thing; they involve different interpretations of the empirical facts, and if we adopt one 'language' we are forced to reject as false certain statements which would be true if we adopted a different 'language'. 'If we accept the sense datum terminology then we must reject the terminology of naive realism; for the two are mutually incompatible'.[1]

It must be said that Ayer himself appears confused on this point. He seems to want to say both that the choice of a sense datum terminology is merely a matter of convenience which makes no difference to what we would ordinarily want to say about what we perceive,[2] and that the sense datum terminology is quite incompatible with our ordinary, Realist, way of talking. Several factors contribute to this confusion. The first, and perhaps most basic, is Ayer's unfortunate use of the term 'sense datum'. We have seen that this should be used as a theory-neutral term, but Ayer explicitly tells us that his 'sense data' exist only when we perceive them,[3] cannot appear to have properties which they do not really have,[4] cannot have qualities they do not appear to have,[5] and are perceived in a sense in which physical objects cannot be perceived.[6] What Ayer, misleadingly, calls 'sense data' I would call 'percepts', and if we are to understand Ayer's position we must read 'percept' for 'sense datum' throughout.

A second factor may be Ayer's sympathy for Phenomenalism. Phenomenalism seems to be a theory which enables us to translate from Idealism to Realism, i.e. enables the Idealist to talk about people perceiving the same object, about objects existing un-

[1] II, p. 48.    [2] II, p. 57.    [3] III, p. 90.
[4] III, p. 89.    [5] III, p. 90 ff.    [6] VI, Ch. 3, §. 3, cf. III, § 3.

THE PHILOSOPHICAL THEORIES OF PERCEPTION 43

perceived, etc., and so seems to be a way of showing that Idealism and Realism are, in the end, different ways of saying the same thing. If that were so, and I will argue that it is not (4.1), Idealism and Realism would not be rival theories, even in the sense of disputing the interpretation of the facts, they would be but alternative and equivalent ways of stating the same theory. Yet it is difficult to see how Ayer can think this so long as he insists that his 'sense datum terminology'—what I prefer to regard as a percept theory of perception—is incompatible with Realism.

This confusion between 'alternative languages' in the sense of different but equivalent ways of saying the same things, and 'alternative languages' in the sense of different and incompatible interpretations of the same empirical facts, affects Ayer's attitude to his theory of perception. One of the most surprising things about his position is that he seems to want to say that if the matter cannot be settled by observation or experiment then, so long as we avoid inconsistency, we can say what we like. To borrow an example from Wisdom:[1] suppose I say that inside my watch is a completely undetectable leprechaun who keeps the machinery moving and in running order. Since no discoverable facts could possibly affect my claim any dispute between me and someone who says there is no leprechaun there is, in Ayer's sense, not a dispute over the facts at all, but merely a dispute as to what terminology or 'language' to use in describing my watch. The choice between my 'leprechaun terminology' and everybody else's 'machinery terminology' will be made on grounds of convenience, misleadingness, and so on. The 'machinery terminology' is obviously preferable both on the grounds of Occam's razor and because the 'leprechaun terminology' might mislead someone into expecting to see a little man in a green hat inside my watch. Nevertheless, Ayer would say, although my terminology, my way of describing the facts, can be castigated as silly, unnecessary, misleading and the like, it cannot be shown to be mistaken; since no proof is possible I can say what I like, so long as I avoid inconsistency.

But this doesn't mean that it makes no difference what we say, for anyone who adopts the 'leprechaun terminology' is committed to the existence of strange, and totally unnecessary, beings, just as anyone who adopts Ayer's 'sense datum terminology' is committed both to the falsity of 'We perceive external objects' and to the

[1] I, Ch. 1.

existence of strange entities which exist only when perceived, are solely and precisely as they appear to be, etc. There is all the difference in the world between 'It doesn't matter what we say because it all amounts to the same thing' and 'It doesn't matter what we say because we can't prove one against the others'. Ayer is so cavalier in his decisions to define 'sense datum' in various ways (it doesn't matter so long as its convenient) only because he doesn't see this difference. For example he claims—with a superficial resemblance to Paul's position—to have argued 'that the solution of the current philosophical problems about sense data depended, not upon our discovering the properties of a strange sort of object, but rather upon our establishing the use of new set of technical terms'.[1] But although Ayer does not, as Moore did, try to *discover* the properties of sense data, what he does do is so *define* the term 'sense datum' that in his usage it must be the name of a special sort of object, a percept. Just as the adoption of the 'leprechaun terminology' does not depend upon our discovering the properties of a strange object but involves the *postulation* of such an object, so the introduction of Ayer's 'sense datum terminology' involves the postulation of a special class of entities.

Ayer is, in effect, defining percepts into existence! There can be no 'doubt that sense data do, in fact, exist . . ., since we have so defined our terms that it must be true that something is being directly apprehended in very case in which it is true to say that anything is in any manner perceived'.[2] We can, if we wish, make it true by definition that whenever we perceive we perceive sense data, but then the question of what sense data are and are like will be an open question. We cannot make it true by definition that whenever we perceive we perceive entities of the very special sort that Ayer describes sense data as being.

I have insisted that a genuine sense datum terminology will be theory-neutral. So if we want to talk about what we perceive without presupposing any particular theory of perception, without interpreting the facts in terms of any one of the three theories, we will have to use not the familiar expressions of ordinary language— for ordinary language is inescapably Realist—but this neutral sense datum terminology. Thus talk about sense data is a way of talking about 'the hard facts' without begging any of the questions at issue between the different theories. In this sense the hard facts,

[1] III, p. 103.  [2] III, p. 90.

## THE PHILOSOPHICAL THEORIES OF PERCEPTION 45

the theory-neutral facts, are facts about sense data. On the other hand Ayer's 'sense datum terminology' is one among several incompatible ways of describing these facts; it is not theory-neutral at all. So it seems that Ayer is trying to have it both ways, to say that the sense datum terminology is only one possible, though perhaps the most convenient, way of describing the facts, and also that the sense datum terminology describes the facts 'as they are in themselves', 'as they really are'.[1] But although Ayer himself may be confused on this point, the fact remains that a sense datum terminology, correctly understood, does provide a theory-neutral way of describing the facts about perception.

No doubt someone will say 'But what are these "theory-neutral" facts? The plain fact is that there is a table—an external object as you would call it—in front of me, and any so-called "theory-neutral" way of describing what I see which makes it possible to deny that I see a table is simply a concealed way of introducing a theory that runs *against* the plain facts. The facts are not theory-neutral; they are what the plain man—a Realist—says they are'. Now it is undeniable that we all naturally take the facts to be what the Realist says they are, and we all ordinarily talk as if they are what the Realist says they are. But—and this is Ayer's point—it is possible to deny what common sense and ordinary language take to be true without this meaning that we have to reject any empirically discovered facts. We will have to interpret and describe these facts in a way very different from the Realist way in which we ordinarily interpret and describe them, but that is a different matter. The facts *are* theory-neutral in the sense that they can be interpreted in accordance with Realism, Idealism or the Causal theory, and no appeal to the facts which does not already involve an interpretation of them in terms of a theory can prove one theory as against the others. The retort that common sense and ordinary language are irreversibly on the side of the Realist interpretation does not prove either that this is the only or that it is the correct interpretation. What it does prove, I think, is that this is the most convenient interpretation.

For Ayer, however, the most convenient interpretation is that of a percept theory. Or, as he would put it, the most convenient terminology for describing the facts is his 'sense datum terminology'. We must consider why he regards this as the preferable

[1] Cf. Austin II, pp. 59–60.

terminology. His reasons seem to be three. First, since the objects of veridical and non-veridical perception are qualitatively alike we might as well have one term to refer to both groups: 'The contention that if these perceptions are not qualitatively indistinguishable the objects perceived must be of the same type, which I saw no reason for accepting when it was treated as a statement of fact, can reasonably be accepted as a rule of language'.[1] Ayer fails to appreciate that this rule of language commits him to the statement of fact he earlier rejected, that if we choose to talk in his 'sense datum terminology' we are committed to the truth of 'There is a class of mental entities, percepts, which we perceive whenever we perceive, whether our perception be "veridical" or not'. Second, Ayer says that his 'sense datum terminology' 'enables us to refer to the contents of our experiences independently of the material things the are taken to present'.[2] This is one advantage of a genuinely theory-neutral sense datum terminology, but we have seen that Ayer's terminology is not theory-neutral in this way. Although it enables us to refer to what we perceive without referring to material things it does this only at the expense of denying that we perceive material things at all. This is a reason for adopting a genuinely theory-neutral sense datum terminology, but it is not a reason for adopting Ayer's terminology. Third, the Realist terminology, Ayer suggests, is inadequate for discussing such things as illusion.[3] But he does not show how it is inadequate. Indeed if these things can be adequately described in Ayer's 'sense datum' terminology, and if, as Ayer argues, that terminology can be explained by reference to our ordinary 'seems' terminology, then these things must be adequately describable in terms of our ordinary 'seems' terminology, without reference to Ayer's 'sense data' at all!

We can, if we wish, adopt a way of talking according to which we always perceive sense-dependent percepts. But it has still to be shown that this way of talking is the correct, or the most preferable, way.

[1] II, p. 25.    [2] II, p. 26.    [3] VI, Ch. 3, § 3.

# CHAPTER 4
# PHENOMENALISM

## 4.1 PHENOMENALISM, IDEALISM AND REALISM

The Idealist maintains that there are no external objects, but he does not want to deny that there are physical objects, things like tables and chairs, houses and fences, which we perceive. And yet there are many things we ordinarily say about physical objects which, on the face of it, the Idealist cannot possibly say. We talk about two people perceiving the same object, of one person perceiving the same object twice, of objects which no one perceives, but two people cannot perceive the same percept, one person cannot perceive the same percept twice, and percepts cannot exist without being perceived. So if the Idealist wants to say that there are physical objects, and that we perceive them, he has to show how this can be said solely in terms of percepts. He does this typically, by maintaining that a physical object is nothing but a collection of percepts, which is to adopt a Phenomenalistic analysis of physical objects. This means that Phenomenalism has come to be thought of as a theory of perception and, like the failure to distinguish sense data from percepts with which it is closely connected, this is a mistake that had done a lot to confuse discussion of the problems of perception. As I understand the term, Phenomenalism is not a theory of perception at all. It is a theory about the relationship between physical objects and sense data, the theory that everything we want to say about physical objects can also be said solely in terms of sense data, the things we 'immediately' perceive.[1] So in so far as 'sense datum' is a theory-neutral term, Phenomenalism will also be neutral as regards the different theories of perception. If we state Phenomenalism in terms of percepts rather than in terms of sense data, or if we identify sense data with percepts, then Phenomenalism will presuppose some non-Realist theory of perception, but there is no reason why we should do this.

Phenomenalism, then, can be distinguished from Idealism, but the two theories tend to go together, since Idealism without Phenomenalism becomes grossly implausible. A Realist can hold

---
[1] These terms will be explained in ch. 11.

that talk about physical objects reduces to talk about sense data, without adding that we always perceive percepts and never perceive external objects. And an Idealist can hold that we always perceive percepts and never perceive external objects, without adding that talk about physical objects is reducible to talk about sense data. The difficulty is that if the Idealist cannot find some way, on his theory, of talking about physical objects, other people and so on, he seems to be committed to Solipsism, the theory that we can know nothing about anything except our own minds and their contents. It seems that the only way in which he can say that we perceive nothing but percepts, and yet continue to talk about physical objects, is by translating talk about physical objects into talk about percepts, which on his theory are identical with sense data. So without Phenomenalism Idealism leads inevitably to Solipsism. This is why the two theories tend to go hand in hand.

Nevertheless the two theories can be distinguished. A disproof of Phenomenalism is not a disproof of Idealism, and a disproof of Idealism is not a disproof of Phenomenalism, although the end of Phenomenalism certainly seems to mean the end of any plausible Idealism. In fact the difficulty is not to show a difference between Phenomenalism and Idealism, but rather to show a difference between Phenomenalistic Idealism and Realism! Suppose that the Phenomenalistic analysis of physical object statements can be carried through, so that everything we want to say in our ordinary Realist way can also be said in sense datum statements. Both Realist and Idealist can now accept all our ordinary statements and beliefs about physical objects. By adopting Phenomenalism the Idealist tries his hardest to be able to say everything the Realist says. And if Phenomenalism works it seems that he can say everything the Realist says. What then is the difference between the two?

The difference lies in their interpretations of our ordinary Realistic statements, statements about external objects. Although both will accept these statements as true they will differ widely in their analysis of the statements, and this will lead to disagreements over the truth value of the further statements that are entailed by those analyses. According to the Idealist 'I see the table I saw yesterday' will mean that I now see an entity (a percept) which is related, as a different part of the same whole, to an entity which I saw yesterday, and no part of the entity I now see will be numerically identical with any part of the entity I saw yesterday. According to the Realist this statement will be false, both in itself and as an analysis of the original statement. Similarly the Idealist will want to accept a statement like 'An entity comes into existence which I open my eyes, and an entity ceases to exist when I close them again', while the Realist will want to reject it. In short the two will differ about the interpretation of sense data, about whether or not they are to be construed as entities in their own right, and about whether physical objects are to be thought of as consisting, in any physical or metaphysical sense, of entities which cannot be perceived on more than one occasion or by more than one person, and which cannot exist unperceived. Finally there will be a difference in their emphasis on the Phenomenalistic translations. The Realist-Phenomenalist looks upon the sense datum translation as an indirect way of talking about external objects—he does not think of the former as replacing the latter—while the Idealist-Phenomenalist regards the external object statements as an indirect way of talking about sense data, which he indentifies with percepts. For the Idealist percepts are the genuine entities and talk about external objects is but a way of referring to percepts, but for the Realist external objects are the genuine entities and talk about sense data is but a way of referring to them.

I have insisted that Phenomenalism is distinct from Idealism. But why have I bothered? Why should anyone who is not an Idealist be interested in Phenomenalism? The answer is that Phenomenalism, if correct, provides an answer to a problem that philosophers have tended to forget in the enthusiasm of their attacks on non-Realist theories. It is usually agreed that it is our perception that provides us with our knowledge of the external world. But our knowledge of the external world includes knowledge of things we have not perceived, things which perhaps no-one has ever perceived. Indeed our knowledge that there is an external world, that objects exist even when no-one perceives them, is knowledge of this kind. How can our perception provide us with knowledge of what we do not perceive? This is the topic we will discuss in the second part of this book, but it can be seen that if Phenomenalism works it may well provide us with an answer to the question, viz. that our knowledge of what we do not perceive is, in the end, equivalent, reducible, to knowledge of what we do perceive. And there is the further hope that if knowledge of sense data is knowledge where we cannot be mistaken we can, via the Phenomenalistic translation, give our knowledge of the external world that absolute certainty that has haunted philosophers since Descartes.

## 4.2 THE TRUTH IN PHENOMENALISM

Phenomenalism has often been attacked, but it always remains tempting. What is the source of this temptation? I think the reason for Phenomenalism's prima facie plausibility is that there is a definite connection between 'exists' in the sense in which we talk of physical and other external objects as existing, and 'is perceptible'. Phenomenalists are particularly concerned to give an account in sense datum terms of what it is for something to exist without being perceived, not only to enable the Idealist to talk about things which are not perceived, but also to answer the Empiricist's problem of how we can know about what we do not perceive. So they point out that to say that something exists, even if it is not perceived, is at least to say that it is perceivible, that it can be perceived, and so long as everyone can perceive a thing in a certain place we say that that thing does exist in that place, even when no one actually does perceive it. In other words to say that something exists is to say that it can be perceived, and vice versa. This is not the end of the Phenomenalistic translation, for the statement about what can be perceived has still to be turned into a statement solely about sense data. But it is a beginning.

It seems indisputable that if, e.g. a stone exists people will be able to perceive it, but this, I think, is a consequence not so much of the fact that the stone *exists* as of the fact that stones are physical objects and physical objects are *sensible* objects.[1] Just as what isn't a man can't be a bachelor so what can't be perceived can't be a stone. This applies as much to hallucinatory stones as it does to real ones; stones, and physical objects generally, are essentially perceptible objects. There are some things, such as atoms and electrons, which we do speak of as existing unperceived and which are not sensible, but the Phenomenalist may be able to allow for their existence in some roundabout way, and we have decided to restrict the discussion to things which are perceptible. So we can

---

[1] Berkeley trades on this point when he says 'To be convinced (that esse is percipi) the reader need only reflect, and try to separate in his own thoughts the being of a *sensible* thing from its being perceived' (I, § 6, my italics).

agree with the Phenomenalist that '$X$ exists independently of our perception' entails '$X$ can be perceived'.

What of the reverse step, from '$X$ can be perceived' to '$X$ really exists'. Obviously there are things—after-images, Macbeth's dagger—which can be perceived but do not exist in this sense, but the Phenomenalist will say that the point about them is that they are, and can be, perceived by only one person. If everyone had perceived the dagger it wouldn't have counted as a hallucination. Or, to rule out the case where everyone perceives what doesn't exist, e.g. because of the presence of a hallucinogenic gas, we might say: if everyone can perceive $X$, and this is not because of some factor not normally present when people perceive, then $X$ exists.

Phenomenalism begins, then, and gains its plausibility from the fact that to say that something, or more accurately some sensible thing, exists, really exists, is equivalent to saying that that thing can be perceived by everyone, given that no special factor affects their perception. But what is the sense of 'can' here?

It can hardly be the 'can' of logical possibility, for it is logically possible for everyone to perceive a window in the wall in front of me, even though there is no window there. If everyone did perceive such a window this would mean that there was a window there,[1] but the logical possibility of everyone's perceiving it does not mean that there *is* a window there. All it means is that it is logically possible that there is a window there. Nor, it seems, can the 'can' be that of empirical possibility, for something might exist without it being empirically possible for anyone to perceive it. There is, for example, a certain piece of coal inside an active blast furnace, but it is not empirically possible for anyone, let alone everyone, to perceive it. Even so we might want to say that although nobody *can* perceive it, anyone and everyone *could* perceive it, if, per impossibile, they managed to survive inside the blast furnace. We shall come back to this suggestion.

The most favoured analysis of 'can' for this, and many another, purpose, is the hypothetical 'would . . . if' analysis. The usual suggestion is that to say that something exists is not just to say a la Berkeley, that it is perceived, but rather to say that it would be perceived if certain conditions were satisfied. Thus 'There is a certain piece of coal inside that blast furnace' is said to be equiva-

---

[1] We might wonder how the window got there, but that would be a different problem.

lent to 'A piece of coal would be perceived if anyone went inside that blast furnace and managed to survive'. However there are several objections to this analysis, and it is worth noticing that they all apply before we try to turn the Phenomenalistic analysis into a sense datum statement, i.e. even when we are still referring to the physical objects.

First, the conditions which need to be stated are extraordinarily complicated. We need to make sure not only that I go inside the blast furnace and manage to survive, but also that I am looking in the right direction, have my eyes open, do not have the coal obscured by some other object (or a hallucination!), am not daydreaming or preoccupied so that I fail to notice it, do not suffer from psychic blindness, etc., etc. This means that we have to add conditions to the original hypothetical until we rule out any possibility of such things preventing us from perceiving what is there to be perceived. But it is at least logically possible that we fail, for no reason at all, to perceive what is there, and no condition can be added which will rule out this possibility, unless we make the hypothetical something like '$X$ will be perceived if we perceive what is there to perceive'. And this incorporates the very notion of existence we are trying to analyse. So long as it is possible for us to fail, for no reason at all, to perceive what is there, '$X$ exists' is not equivalant to anything of the form '$X$ would be perceived if . . .', unless this latter incorporates the claim that $X$ exists.

If this objection, resting on a mere logical possibility, is not found convincing other objections can be added. There may well be no limit to the number of things which might go wrong, which would prevent us from perceiving what is there to be perceived. In that case no finite list of conditions will make the hypothetical equivalent to '$X$ exists'. Again, some of the conditions which have to be added presuppose the notion of real existence. What does psychic blindness consist in if not being unable, for some psychological reason, to perceive what is there? Similarly the condition that we look in a certain place involves the notion of existence, that there exists a place where the object will be perceived. No doubt the Phenomenalist will try to eliminate this at the next step, when he reduces the statement about what can be perceived to a statement about sense data, but it is not clear how he can handle a bald existential like 'My birth certificate exists'. Phenomenalists usually concentrate on statements like 'There is a table in the next room' which suggest the conditions that have to be satisfied if the object is to be perceived. But it seems that the only analysis he can give of 'My birth certificate exists' is 'There exists some place such that if you go there you will see my birth certificate'. And even if the Phenomenalist can, as seems unlikely, specify in sense datum terms conditions which rule out the possibilities of absent-mindedness, looking in the wrong direction, psychic blindness, etc., there is the further objection that he is involved in an infinite regress. For we have to be sure that the failure to perceive sense data indicative of absent-mindedness, looking in the wrong direction, psychic blindness, etc., is not itself due to absent-mindedness, looking in the wrong direction, psychic blindness, etc. If this possibility is ruled out by reference to further sense data the same problem arises again, and so on *ad infinitum*.

Finally there is the 'Dr Crippen' objection. Dr Crippen murdered his wife at a certain place at a certain time, but it does not follow that if we had been there at that time we would have perceived him murdering his wife. Had we been there he would, presumably, have postponed his deed till some more opportune occasion. So 'Dr Crippen murdered his wife' is not equivalent to 'If we had been at the appropriate place at the appropriate time we would have seen Dr Crippen murder his wife'. I think this argument is quite correct, but we may feel unhappy about it. Surely, we want to say, there is a sense in which we would (or, perhaps, could) have seen the crime had we been there, simply because the crime did occur. This gives the game away. We want to say that we can see a table in the next room, could see Dr Crippen murder his wife, could see the coal in the blast furnace, *just because* there is a table in the next room, because he did murder her, because the coal is in the blast furnace. The sense in which it is true that what exists can be perceived is one which presupposes the notion of real existence itself.

This becomes clearer, if we return to the 'can' of empirical possibility. It is, in a sense, empirically possible to see the coal in the blast furnace, i.e. we could see it if we managed to survive. But it is not, in that sense, empirically possible to see a window in the wall in front of me, or to see a pink elephant. Why is it possible for me to see the coal? Because it is there. Why is it impossible for me to see the window, or a pink elephant? Because there is no window there, no pink elephant.

To sum up: Phenomenalism gains its plausibility from an equivalence between '*X* really exists' and '*X* can (or could) be perceived by everyone, without this being the result of some special factor affecting their perception'. But this equivalence cannot be used to analyse the notion of real existence because the sense of 'can be perceived' involved is one which itself presupposes the very notion of real existence. Only things which exist can be perceived in this sense of 'can be perceived'. In fact Phenomenalists have often maintained the direct reverse of the truth. In trying to avoid the epistemological problem that leads to a Berkelean Idealism, the problem of how we can know about what we do not perceive, they suggest that it is true that something exists unperceived only in so far as it is true that we can, given certain conditions, perceive that thing.[1] But the truth is quite the opposite. It is true that we can, given those conditions, perceive that thing only in so far as it is true that that thing does exist.

It is worth repeating that the sense in which this equivalence holds is not that in which 'can' can be translated by 'would . . . if'. The difficulties of 'can', the toad in the bottom of the beer mug, are notorious. It is, in particular, easy to confuse 'could . . . if' with 'would . . . if' (we are, for example, uncertain whether to say that we would have seen Dr Crippen murdering his wife, or merely that we could have seen it) and I think it is this confusion which makes the 'would . . . if' analysis of 'can' so tempting to many Phenomenalists. At first we want to say that if there is a table in the next room then if we were to go there we would see it, and vice versa, but, unfortunately, we might go there and fail to see it for any number of reasons, or even for no reason at all. What is true is that if we went there we *could* see it, and this remains true whether we actually see it or not. Similarly to say that my birth certificate exists is not to say that I will see it if certain conditions are satisfied, for I can say and even know that it exists without saying or knowing what those conditions might be. Rather it is to say that it *can* be perceived.

## 4.3 OBJECTIONS TO PHENOMENALISM

The argument of the last section seems to me to show not only what the truth behind Phenomenalism is, and why it is so persistently plausible, but also to show that, in the last analysis, it cannot

[1] Cf. Ayer I, p. 145.

## PHENOMENALISM

work. However there are many other arguments that have been brought against the theory and in this section I want to state as briefly as possible those which seem to me either the most frequent or the most convincing. I begin with three arguments which do not seem to me to be successful.

(1) It is argued that a sense datum statement can never entail an external object statement, that no matter how many sense data can be perceived it does not follow that an external object exists.[1] This is sometimes put in the form that external object statements are infinitely verifiable; no matter what we perceive there is always the possibility that something will go wrong, no matter how consistent our perception of it there is always the possibility that it doesn't exist after all. I am convinced that my wife exists, but it is logically, even empirically possible that she does not, despite all my perception of her. I may wake up to find myself surrounded by eminent brain surgeons who inform me that there is no such woman, that the whole thing is simply a result of their monkeying around with the perceptual and memory centres in my brain.

All that this argument establishes is that it never follows from what one person perceives that some particular thing really exists. But the Phenomenalist is not trying to establish an equivalence between statements about what really exists and statements about what one person happens to perceive. The equivalence is supposed to be between statements about what really exists and statements about what *can* be perceived. If everyone can perceive this woman, and this is not the result of some special factor affecting their perception, then she does exist. This reply is open to the further objection about the sense of 'can' involved, but that is a different objection.

Notice, too, that external object statements are not infinitely verifiable. It follows from the fact that so many people have so often perceived parts of the island we call Great Britain that this island does exist. The existence of Great Britain has been conclusively verified (if it doesn't exist what does?). Of course I may be mistaken as to whether all these people have perceived it, but the point is not that we cannot establish, with logical conclusiveness, that something really exists, but that if we are to do this we must refer to the perception of more than one person. We will return to this in objection 11.

[1] E.g. Ayer IV, pp. 134–8.

(2) Conversely it is argued that external object statements can never entail a sense datum statement, because even though the object exists it does not necessarily follow that these particular sense data will be perceived.[1] Once again the argument forgets that the Phenomenalist is drawing an equivalence between external object statements and statements about what *can* be perceived, not about what actually is perceived. '$X$ exists' is held to be equivalent, not to 'Such and such sense data are perceived', but to 'Such and such sense data can be perceived'. Moreover the argument often involves a confusion between truth and verification. Urmson argues[2] 'It is clearly not a necessary condition' (of the truth of the external object statement) 'that those particular sense datum statements should have been the verified ones; others would have done as well'. True enough, but what is supposed to be the necessary condition is not that the sense datum statements be *verified*, but that they be *true*. The best way of verifying 'Such and such sense data can be perceived' is to perceive the relevant sense data, but the statement may be true even if it is not verified in this way. So '$X$ exists' may entail 'Such and such sense data can be perceived', and vice versa, even though those sense data are not actually perceived.

(3) There is also an argument that external object statements cannot be translated into sense datum statements because there cannot be a sense datum language in the first place. A discussion of this 'private language argument', the argument that there cannot be a language with terms referring to private objects, would take us too far from our present topic. All I will say is that the most the argument seems to me to prove is that there cannot be a sense datum language unless we already have an external object language. This would seem to count against the Idealist who hopes to use the Phenomenalistic analysis to replace our ordinary external object talk, but not against the Realist who simply regards the two as interchangeable. And perhaps even the Idealist can say that all this means is that we have, unfortunately, to approach the truth via false assumptions, which we can then dispense with.

The next two arguments count not against Phenomenalism as such but against the Phenomenalistic version of Idealism. The difficulties arise not in the translation from statements about ex-

[1] Cf. Ayer IV, pp. 138–40.      [2] I, p. 157.

ternal objects to statements about sense data, but in the identification of sense data with percepts:

(4) First there is a group of objections which arise from the Idealist's analysis of our ordinary statements about our perception. We have seen that this analysis involves certain claims which go against our common sense beliefs, such as the claim that when I begin to see there springs into existence an entity which did not exist before, or the claim that I am at this moment seeing some thing which is in no respect numerically identical with what I was seeing two minutes ago, and so on. The objections take many forms: it has been argued that the Phenomenalist is committed to saying that what I see is not spatially related to what you see in the way that the typewriter I now see is spatially related to the telephone I now see (and similarly for time);[1] that the Phenomenalist is committed to saying that perception of public objects involves, and even more oddly consists of, perception of private objects;[2] that, in the last analysis, the Phenomenalist make the existence of external objects a conditional or hypothetical matter;[3] or simply that the Phenomenalist accepts such statements as 'The table exists' only by giving them a Pickwickian sense.[4] But it is not usually noticed that these are criticisms of the Idealist interpretation of Phenomenalism, and not of Phenomenalism as such. Indeed some criticism of this sort—that even a Phenomenalistic Idealism is not entirely consonant with common sense—*must* be possible if there is to be any difference between a Phenomenalistic Idealism and a common sense Realism.

Perhaps the most striking of these arguments is that, despite his Phenomenalism, the Idealist is still committed to the view that there could be no existence without minds, that the existence of matter depends on the existence of perceivers. 'If the Phenomenalist is correct then before there were minds having sense-impressions, there was just nothing. You can say "There were physical objects", but if this only means that there were unfulfilled possibilities of having sense-impressions this does not contradict the statement that there was nothing'.[5] If we do identify sense data with percepts, thinking of these as the genuine constituents of the world from which what we call 'external objects' can be constructed,

---

[1] Cf. Armstrong I, pp. 62–7.   [2] Cf. Hirst I, pp. 94–7.
[3] Cf. Berlin I and Ayer's reply VI, pp. 120 ff; Armstrong I, pp. 53–6.
[4] Cf. Moore II, pp. 190 ff.   [5] Armstrong I, p. 55.

then we are committed to the startling view that 'Physical objects existed before there were perceivers' is consistent with, and even entails, 'Nothing existed before there were perceivers'. For to say that physical objects existed before there were perceivers is simply to say that certain things (percepts) which could have existed did not. Obviously this is quite contrary to common sense, although, as Armstrong admits, it does not mean that an Idealistic Phenomenalism must be mistaken. But a Realist Phenomenalist is not committed to the oddity. He says that things did exist before there were perceivers, although he adds that this is equivalent to saying that parts or aspects of those things could have been perceived had certain conditions, e.g. the existence of perceivers, been satisfied. To say that something can be perceived if something else is perceived—which is the form of the Phenomenalist's sense datum statements—is not to say that anything is perceived or even that any perceivers exist, so its truth is quite compatible with the non-existence of perceivers. It is only the Idealist, who identifies sense data with percepts and claims that percepts are the only things that do exist and that everything else which is said to exist is made up of them, who is forced to make the existence of matter depend upon the existence of mind.

(5) There is a rather different objection that the Phenomenalist is forced to reject, or at best make nonsense of, our ordinary causal, and similar kinds of, explanations.[1] The room I am now in is supported by foundations which no one perceives. What I now perceive is actual but the foundations, according to the Phenomenalist, are mere unfulfilled possibilities. How can actualities depend upon unfulfilled possibilities?

In a way this question is as unfair as the older 'Would you have me say that we eat and drink percepts, and are clothed in percepts?'[2] The Phenomenalist does not deny that the foundations exist any more than he denies that there are clothes and food. But the Idealist allows that the foundations exist only in the sense that percepts could exist given certain conditions, i.e. he reduces the actual existence of the foundations to the possible, not actual, existence of private mental entities. So he does seem committed to the view that this room is supported by a collection of unful-

[1] Cf. Stout I; Hardie I; Ayer IV, pp. 143–50.
[2] Cf. Berkeley I, sect. 38.

filled possibilities, and this seems perilously close to nonsense. The oddity does not arise for the Realist Phenomenalist. He insists that the foundations do exist, and although he adds that this amounts to saying that certain sense data can be perceived, he does not mean that saying the foundations exist is nothing more than saying certain private mental entities which do not exist could exist. All he means is that, whether they are perceived or not, certain parts or aspects of the foundations could be perceived.

One way out of the Idealist's difficulty might be to say that questions of cause and effect, etc., apply only at the level of external objects, that these notions cannot be applied to percepts.[1] But how can this restriction be justified, except as an *ad hoc* device to avoid oddities? Why should we not ask what causes our percepts? Surely we want to ask, as Locke did, why our percepts are as they are, why certain visual ones go with certain tactual ones, why different people perceive similar ones? It may that we cannot answer these questions, except by means of a Berkelean *deus ex machina*. But the only reason for saying we shouldn't even ask them seems to be that this isn't convenient! The Idealist must be committed to odd and implausible—though not necessarily false— views about the nature and status of causal and similar types of explanation.

The following arguments against Phenomenalism seem to me to be successful:

(6) The first is the argument we have already considered, that in the sense in which '$X$ really exists' is equivalent to '$X$ can be perceived', 'can be perceived' itself relies upon the notion of $X$'s really existing. What this means, in effect, is that the notion of a possible sense datum, which is indispensible for the Phenomenalistic analysis, incorporates the notion of real existence which is to be analysed. The Idealist has to refer to possible sense data if he is to explain not only what it is for something to exist unperceived, but also what it is for two people to perceive the same thing, or even for one person to perceive the same thing on different occasions. What makes it true that Jones and Smith are perceiving the same pen is not that they are perceiving similar sense data, for pens are mass-produced and they may well perceive similar sense data when they are looking at quite different objects. There is even

[1] Cf. Ayer IV, pp. 146-50.

the possibility that the surroundings in which they perceive the pen are qualitatively the same, even though what they perceive are thousands of miles apart. What makes it true that they are perceiving the same pen and not different pens in similar surroundings is not that what one sees is pretty much like what the other sees—for if one looks through distorting glass the differences might well be great—but that they are both looking at the same part of space. And that they are both looking at the same part of space can only be guaranteed, in sense datum terms, by talking about what other sense data are possible if they each do different things, such as looking around them, moving away in various directions, etc. I shall later argue that even this does not provide a Phenomenalistic analysis of what it is for two people to perceive the same thing, but at any rate it is obvious that the analysis cannot even hope to succeed unless it refers to possible sense data. And the present argument is that by a 'possible sense datum' we mean not one that is logically possible (for any sense datum that is describable without self-contradiction is logically possible), nor one that is empirically possible (for any number of hallucinations are empirically possible) nor one that will be perceived if certain conditions are satisfied (for we cannot specify conditions that guarantee that what exists will be perceived), but one that is a sense datum of something that does really exist. The Phenomenalistic analysis is, in the end, circular.

(7) Even if the Phenomenalist does not adopt a 'would . . . if' analysis of 'can' most of his sense datum statements—all of those that assert that something exists in a certain place—will be hypothetical statements, of the form 'If such and such sense data are perceived other sense data can be perceived'. How are we to tell whether these hypotheticals are true? The difficulties of the verification of hypotheticals, particularly counter-factual hypotheticals, are well known. The logician avoids them by saying that so far as he is concerned a hypothetical is true so long as it does not lead from truth to falsity, so long as the apodosis is not false when the protasis is true. This is not how we would ordinarily determine the truth of a hypothetical, and it is not how the Phenomenalist wants to determine the truth of his hypotheticals. On this interpretation 'If I perceive Taj Mahal-like sense data then I can perceive Albert Memorial-like sense data' is true so long as I do not perceive Taj Mahal-like sense data, whereas the Phenomenalist

PHENOMENALISM                    61

wants this statement to be true if and only if the Albert Memorial is in, on or near the Taj Mahal.

I think that the question of whether a particular hypothetical is true or false is, ordinarily, the question of whether some categorical—often universal—statement, in which the hypothetical is, as we might say, grounded, is true or false. That is, when a person claims that something of the form 'If $p$ then $q$' is true we ask him to justify his statement and show that it is true by indicating some true categorical statement which establishes the truth of the hypothetical (we need not go into the difficult question of how it does this). The hypothetical is grounded in the categorical in the sense that it is true or false (or a matter of opinion) in so far as the categorical is true or false (or a matter of opinion). The categorical may be the statement of some natural law or empirical generalization but it need not be. The hypothetical 'If I let go of this book it will fall to the floor' is grounded in some such categorical as 'All unsupported bodies fall', and the hypothetical 'If Hitler had not invaded Russia he would have conquered Britain' is grounded in some such categorical as 'Hitler's undivided forces were stronger than those of Britain'. The important point is that an ungrounded hypothetical, one without a categorical on which its truth value depends, is not only impossible to verify but actually lacking in truth value.

The Phenomenalist's sense datum hypotheticals will, presumably, be grounded in the categoricals to which they are supposed to be equivalent. 'If I perceive next room-like sense data I can perceive table-like sense data' will depend for its truth value on 'There is a table in the next room'. So if the sense datum statement is to be true it requires and presupposes, and cannot replace, the external statement which it is supposed to translate. The Phenomenalist may say that what this shows is that the categorical is needed as well as the hypothetical and he is, after all, not denying or rejecting the categorical, but maintaining only that one is a translation of the other. But there is more to the argument than this. First of all it shows that it is quite impossible for the hypothetical to be used instead of the categorical, as the Idealist wishes. We cannot say that the categorical is just a construction on the true facts which are described by the hypothetical, for the argument shows that the truth is the other way about. And secondly it shows that the sense datum hypothetical cannot be a complete translation of the categorical. No matter how we state the hypothetical it always presupposes some categorical statement in which it is grounded, so there must always be some categorical element required by the hypothetical and not included in it. We may be able to deduce, logically, one from the other, but this is possible only because one presupposes the other. It is this point, no doubt, which explains both the feeling that even if the Phenomenalistic translation is entailed by and entails the ordinary external object statement, it does not contain the full meaning of that statement, and the feeling that the Phenomenalist has done something strange to the solid enduring existence we ordinarily accord to such things as irons and inkwells.

In a way this objection is even more devastating than the previous one. It amounts to saying that the Phenomenalist's translation presupposes not just the notion of real existence, but everything which is supposed to be translated. But in another way it may not be such a strong objection. For the Phenomenalist may be prepared to accept the point, saying that his aim is not to replace the external object categorical by a sense datum hypothetical, nor to say that the sense datum hypothetical gets closer to the facts on which external object talk is based, nor even to offer the hypothetical as a full translation of the categorical, but simply to show that the hypothetical entails and is entailed by the categorical. Even so it is doubtful whether the Idealist can be satisfied with this, and in fact most Phenomenalists have wanted to say more.

(8) There is a familiar argument that the Phenomenalist's analysis of physical objects is circular. The claim is that physical objects 'consist of' sense data, but we cannot explain which sense data 'make up' the object except by referring to them as sense data of that object. We have to refer to the object in order to explain which sense data it consists of. This is a consequence of the general point that objects are identified by reference to their spatio-temporal location. We cannot identify particular objects in purely sense datum terms; we can do this only by reference to a fixed spatio-temporal framework which is itself established by reference to external physical objects.[1]

There are two ways in which people have tried to establish that a sense datum is a sense datum of one particular object. The first works by reference to the 'sensory setting'. We try to identify the

[1] Cf. Strawson I, Ch. 1.

object in sense datum terms by describing what sense data can be perceived if we look around the place where the first sense datum is perceived. There are two difficulties here. The first is that we may never know when we have mentioned enough sense data to distinguish, e.g. this particular room and this particular table from all other rooms and tables, and the amount and type of sense data that have to be mentioned will vary from room to room and table to table. And secondly it is always an empirical question whether or not there is something else qualititatively identical with this room and this table, so the connection between 'There is a table in the next room' and the sense datum statement can never be one of logical entailment. It may follow from the fact that I perceive sense data, $a$, $b$, $c$ that I am perceiving external object $A$ and not something else like $A$, but this follows only because there is not something somewhere else which is exactly like $A$ in these respects, so that I could perceive the sense data I do and yet be perceiving not $A$ but this other thing. And that there is no such thing exactly like $A$ in these respects is always an empirical and never a logical matter, so it cannot follow logically from the fact that I perceive these sense data that I am perceiving $A$ and not something else. This objection holds no matter how far we specify sense data of $A$ and its surroundings. Even if we specify all possible sense data—presumably an infinite number—there is still the logical possibility that the world is symmetrical such that one could perceive qualitatively identical sense data in precisely the same order by starting out from numerically distinct objects in numerically different places.

The second way of trying to specify in sense datum terms that a particular object is being perceived is by reference to the 'sensory route' that has to be traversed in getting from the place now perceived to the place where that object is. On this account the sense datum translation for e.g. 'There is a colony of penguins at the South Pole' will be enormously complicated, referring to all sorts of things—everything that is to be perceived on the way from here to the Pole—about which I know, think and care nothing and which, therefore, do not seem to belong in any analysis of my original statement. Moreover the 'sensory route' method is open to precisely the same objections as the 'sensory setting' method, for we will need to establish that the sense data mentioned in the sensory route story are sense data to be perceived on the way to the South Pole, and not sense data to be perceived on the qualitatively similar way to somewhere else. And this raises just the difficulties of trying to establish that these are sense data of the table in the next room.

These difficulties are familiar but they lead to a stronger objection which is not well-known. No sense datum statement which is entailed by e.g. 'There is a table in the next room' could ever be equivalent to it, because it would not entail it in turn. If a sense datum statement is to entail 'There is a table in the next room' it will have to be sufficiently detailed to distinguish this table and this room from all other tables and rooms, and it will be able to do this only by saying what this table and this room are like. But 'There is a table in the next room' tells us nothing about what the table and room are like; table and room could be very different and yet 'There is a table in the next room' still be true. In trying to establish, in sense datum terms, that there is a table in the next room, as opposed to there being a table in some room somewhere, the sense datum statement must, of necessity, go beyond what is in the original external object statement. The point is that we can identify external objects by referring to them, without saying what they are like. But since sense datum statements cannot refer to external objects we cannot identify external objects in sense datum terms without saying what they are like, and so going beyond the original external object statement. No sense datum statement which entails 'There is a table in the next room', as opposed to there being a table in some room somewhere, is entailed by that statement. And vice versa.

The Phenomenalist may now say that all he needs is an empirical, and not a logical, equivalence between 'There is a table in the next room' and the sense datum statement. It is not clear what this modification does to Phenomenalism, but the Phenomenalist is still faced with overwhelming practical difficulties. There may be some description of the table and the room next door which might be put into sense datum terms and which is sufficient to distinguish this table and this room from all other tables and rooms. But I have no idea what this description is, and I could never be sure that I had got the correct identifying description until I had examined all other tables and rooms. Even if all external object statements have sense datum statements to which they are equivalent as a matter of fact, it seems that we can never know what these sense datum equivalents are. There may be Phenomenalistic 'translations', in this sense, for our ordinary external object statements, but we can never provide them!

(9) When the Phenomenalist offers something of the form 'If such and such sense data are perceived such and such other sense data can be perceived' as a translation for an external object statement he tacitly assumes that the perceiver remains in one spot, or at the most moves only a small distance. For it might be the case that we first perceive sense data as of the next room and then perceive sense data as of the table even though there is no table in the next room, because, between perceiving the first and second sense data, we are moved from the next room to some other place where the table is. How can this assumption be stated in sense datum terms? We will not be able to rule out the possibility of movement by referring to other visual, tactual or kinaesthetic sense data for it is always possible for us to move or be moved without having any of these visual, tactual or kinaesthetic sense data. The only way of ruling out this possibility is to refer to the sense data other perceivers could perceive if they were to watch us. We then have to ensure that they are not the victims of instantaneous transportation. And so on, *ad infinitum*.

(10) The Phenomenalist has difficulties over the notion of a perceiver. He will have to avoid any reference to perceivers who are to be thought of as physical or external objects, as involving more than sense data. Ayer[1] has suggested that this might be done by phrasing the sense datum statements 'impersonally', by saying, e.g. 'If sense data as of the next room are perceived sense data as of the table can be perceived'. But it is clear that this statement still presupposes a reference to a perceiver, to one perceiver who perceives both room-like and table-like sense data. It may be true both that I perceive sense data as of this room and that some London bus conductor perceives sense data as of the Albert Memorial, without it being true that this room is in, on or near the Albert Memorial. The Phenomenalist will have to produce some Neutral Monist account of what a perceiver is, i.e. a definition of a perceiver solely in terms of sense data. There is not space to go into the difficulties of such an account, but it seems to me that a Phenomenalistic analysis of perceivers is, if anything, less likely to succeed than a Phenomenalistic analysis of external objects.

[1] IV, pp. 162–3.

(11) There is a final point that might be made against the Phenomenalist, although it is not a conclusive theoretical objection. One of the main motives for Phenomenalism, whether Realist or Idealist, is that it can cash everything in terms of what is, or can be, perceived, and so enables us to avoid the problem of how we can know about what cannot be perceived. But we have seen that a Phenomenalistic translation has no hope of succeeding unless it refers to sense data perceived or perceptible by more than one perceiver. It never follows logically from what I perceive that other people perceive, or can perceive, certain things, so the Phenomenalist has to refer if not to things which cannot be perceived at least to things which any particular individual does not and cannot perceive. So we are left with a form of the problem that Phenomenalism was designed to avoid. We are left with the problem that what we claim to know involves a reference to what we, individually, do not and cannot perceive.

## 4.4 'PRAGMATIC' PHENOMENALISM

These arguments show that the Phenomenalist's attempt to find sense datum statements that entail and are entailed by external object statements must fail. They also show that a weakened Phenomenalism that claims only an empirical implication from external object statement to sense datum statement, and vice versa, fails as well, whatever the point or status of such a Phenomenalism might be. However these objections might be avoided if we adopted a 'probabilistic' or 'pragmatic' Phenomenalism which explicitly rejects the possibilty of mutual entailments or implications and claims merely that the external object statement is, for all practical purposes, equivalent to a statement that, given certain conditions, various sense data are likely to be perceived.[1] Various sense data are perceived and so we say things like 'There is a mouse in the corner' or 'There is something supporting this floor'; to say these things is just to say that certain other sense data of various appropriate kinds are liable or likely to be perceived by perceivers who get themselves into certain places, this in turn being explained in terms of sense data that are liable or likely to be perceived. If no-one perceives the appropriate sense data we reject the external object statement; if some do and some do not perceive them then we may be in doubt as to the truth of the

[1] Ayer IV, sects. 3, 7; Lewis I.

external object statement; but if everyone or most people or even just several people do perceive the appropriate sense data the external object statement is taken as true.

This attentuated Phenomenalism is difficult to prove false, if only because of the essentially vague way in which it is stated. The relation held to hold between sense datum and external object statements is so indeterminate as to be difficult to examine, and the theory rests on the obvious fact that the external object statement and the sense datum statement will be in much the same position as regards evidence, verifiability and content. Evidence for and proof of one will be evidence for and proof of the other, and to the same extent, and, for all practical purposes, the one conveys as much information as the other. Even so objections can be raised. Talk about 'probable' and 'likely' sense data may enable the Phenomenalist to escape the difficulties of the notion of 'possible' sense data, but perhaps the same objections arise. In what sense are these sense data probable or likely? Why are they probable or likely? Surely their probability and likelihood can only be established, even made sense of, by reference to the empirical facts, the facts as stated in the external object statement which is to be translated. Does not the sense datum statement presuppose and rely upon the external object statement? Nor do I see how a 'Pragmatic' Phenomenalism can overcome the impossibility of translating identifying references to external objects into sense datum terms (objection 8). Even this drastically weakened Phenomenalism does not seem to escape the fundamental objections to the theory.

CHAPTER 5

# THE ARGUMENTS TO SENSE-DEPENDENCE

## 5.1 PRIMARY QUALITIES AND SECONDARY QUALITIES

If it is impossible to provide a Phenomenalistic analysis of statements about physical objects, it follows that the Idealist cannot accommodate the views of common sense about what we perceive. It he cannot show that everything we ordinarily say about what we perceive can also be said in terms of percepts, he cannot show that these common sense views are consistent with the assumption that there are no external objects. The Idealist theory will have to be a strict or pure Idealism which speaks only of percepts, and which runs directly counter to our ordinary views about the world around us.

What arguments are there for such a theory? Or more widely, what arguments are there for the view, accepted by both the Idealist and the Causal theorist, that the things we perceive are sense-dependent, exist only in our perception of them. We have seen (3.2) that if we are not careful about how we describe what we perceive, it may be easy to take it for granted that whenever we perceive we perceive special entities which exist only in so far as we perceive them. But once we think about it we can see that this conclusion stands in need of proof. Perhaps the most common argument is the so-called Argument from Illusion. We will consider it in the next chapter. In this chapter we will discuss the arguments about 'secondary qualities', and the Argument from Sensations.

It is often held that certain qualities which we perceive, the secondary qualities as they are called, depend for their existence on our perception of them. Although this does not in itself establish that everything we perceive exists only in so far as we perceive it, this conclusion does follow if we agree that we perceive the primary qualities only through our perception of the secondary qualities. We may still want to say that there are instances of the primary qualities existing independently of our perception, but it will not be these independent primary qualities which we are aware of in our perception. Rather the primary qualities we perceive, and which are in our perception indissolubly linked with the sense-dependent secondary qualities, will be, at best, sense-dependent representations of the independent primary qualities. This, roughly, was Locke's view. Or one might argue, as Berkeley does in effect, that the secondary qualities are sense-dependent, and that there is no genuine difference between the primary qualities and the secondary qualities, which means that all sensible qualities will be sense-dependent. So in either of these two ways the arguments for the sense-dependence of secondary qualities might be used to establish the sense-dependence of everything we perceive.

The primary qualities are shape, size, extension, position, solidity, impenetrability and rigidity. To these are sometimes added number, which hardly counts as a quality, and motion and rest, which are simply aspects of position. The secondary qualities are colour, taste, smell, temperature, texture and sound. What are the differences between these two groups? Do they justify our saying that either or both are sense-dependent?

The first thing that is said about the secondary qualities is that they vary with the conditions of observations in a way that the primary qualities do not. If I look at a white object through red glass I see red not white; from a distance a loud sound is heard as soft, if we have a cold things lose their smell, etc. On the other hand an object's size and shape do not vary in this way, so it seems natural to conclude that the former qualities depend on the perceiver in a way that the primary qualities do not. To this Berkeley replied that the primary qualities do vary according to the conditions of observation—a distant object looks small and a tilted penny looks elliptical just as the white wall looks red and the distant sound sounds soft.

Even so there is an important difference here which Berkeley's argument obscures. With both primary and secondary qualities an object can appear different from what it really is, but the secondary qualities are *appearance-determined*, i.e. the nature of the quality is determined by how it, the quality, appears.[1] There may not be much point or even sense in asking 'What shape is this square?' or saying 'This red is red' but we can ask 'What shape is this shape?' and say 'This colour is red'. Now if it makes sense to ask which quality a quality is, or to say that a quality is a particular quality, it

---

[1] 'Appears' here has what I will call its 'resemblance' sense, cf. 6.3.

should make sense to say that a quality appears to be other than it is. There seems nothing odd about saying that a shape appears to be elliptical but is, in fact, circular, nor in saying that a circle appears elliptical. But what of 'This colours appears to be red but is blue' or 'This (expanse of) blue appears red'? Surely these only make sense if we take them to mean 'The colour *of this thing* appears to be red but is blue' or 'This blue *object* appears red'. The thing I see can appear a different colour from what it is, but surely the colour I see cannot appear different from what it is? For a colour is determined by its look; if a colour looks blue then it, the colour, is blue, which is not to say that if a thing looks blue it, the thing, is blue.

To bring this out in another way: if I look through blue glass at something which is white I will not perceive any white at all, but only blue; the colour I see is not the colour of the object but the colour the object appears under such circumstances. Three things are true: *the thing* I see *is* white; *the thing* I see *looks* blue; and *the colour* I see *is* blue, because it, the colour, looks blue. This is what I mean by calling colour an 'appearance-determined quality'; what quality I perceive is determined by how what I perceive appears.

Shape, on the other hand is not an appearance-determined quality. There is no oddity in saying 'This shape appears elliptical but is circular' or 'This circle appears elliptical' because what shape a shape is is determined not by how it appears but by reference to such things as rulers and protractors. If a thing looks as though measurement would show it to be elliptical, when in fact it shows it to be circular, then not only is the thing circular, but also the shape which I see is a circular, though elliptical-looking, shape. Three things are true: *the thing* I see *is* circular; *the thing* I see *looks* elliptical; and *the shape* I see *is* circular, a circular shape which, thanks to the laws of perspective, looks elliptical.

If the test for what colour a colour is were not how it appears but, as with shapes, some form of measurement, via a light-meter for example, i.e. if the final test for what colour a colour is were what the light-meter registered, then we could say, as now we cannot, that a colour looks different from what it really is, i.e. we could say such things as 'The colour looks green but, as the light-meter shows, is really blue', just as we can now say 'The shape looks elliptical but, as measurement shows, it is really circular'. In that case colours would not be appearance-determined.

Sounds are more complicated. Even if my ears are blocked with cotton wool so that a loud and piercing sound sounds soft and muffled we want to say that the sound I hear is loud and piercing, just as we say that the shape I see is circular, although it looks elliptical. But on the other hand we want to say that I hear it as soft and muffled just as, through the blue glass, I perceive the white object as blue, when we could hardly say that I perceive the circular penny as elliptical.[1] So it is not clear whether sounds are appearance-determined qualities or not. The truth is, I think, that sounds are not qualities at all but things, external objects, in their own right. If I say that the colour I see is, really is, there to be seen then I necessarily commit myself to saying that there is, really is, something there which possesses that colour. There seems to be a contradiction in the notion of a colour's existing without there being anything which possesses, or at least appears to possess, that colour. But, even if it never happens, there is certainly no contradiction in the notion of a sound's existing without there being anything which makes that sound. When I say that the sound I hear is, really is, there to be heard I do not necessarily commit myself to saying that there is, really is, something there which makes the sound. Sounds, then, are items in their own right, but like physical objects they possess various qualities, of being sharp, flat, shrill, piercing, muffled, etc., and it is these qualities of sounds which are appearance-determined. And, to be quite accurate, it is these qualities of sounds which are the secondary qualities, rather than sounds as such. All this can be also said, I think, about tastes and smells.

Finally, notice that with appearance-determined qualities the one thing cannot appear, in this sense, to be two different things at once. A thing cannot look white and look blue at the same time in the way that the tilted penny does look both circular and elliptical. We might say of the thing that looks blue through blue glass

[1] In fact the 'perceive as' expression seems appropriate only where there is the possibility that the quality I perceive is different from the quality actually possessed by the object, as with appearance-determined qualities or when, for example, a match-box seen in a distorting mirror looks oval. In the match-box example, although not in the tilted penny example, we want to say that the shape I see is not the shape of the object. Rather it is the shape of the object's mirror image.

that it *looks like* a white thing, meaning that it looks as we would expect a white thing to look in the circumstances. But we cannot say that it looks white, *tout court*, as we can say both that the penny looks circular and that the penny looks elliptical.

There is, then, a genuine difference between the primary qualities and the secondary qualities in that the later are, while the former are not, appearance-determined. So when an object appears to have a different secondary quality from the one it really has, the quality perceived cannot be the quality possessed by the object. But when an object appears to have a different primary quality from what it really has the quality perceived need not be different from the quality possessed by the object.[1]

A second difference claimed between primary qualities and secondary qualities is that the latter are closely linked with the particular senses we possess, in a way that the latter are not. This is certainly true. For a start, colours, sounds, etc., are what have been called the 'proper objects' of their respective senses. This means that we cannot perceive those particular qualities except via the particular sense, i.e. we cannot perceive colours except via sight, we cannot perceive sounds except via hearing, and so on. Moreover, colours, sounds, etc., are also what have been called the 'tautologous objects' of their respective senses. This means that we cannot perceive via these particular senses without perceiving those particular qualities, i.e. we cannot see without seeing colours, we cannot hear without hearing sounds, and so on.

Primary qualities, on the other hand, do not seem to be necessarily linked with the particular senses in this way. Except for the trio of solidity, rigidity and impenetrability, all of the primary qualities can be perceived by more than one sense—they can all be perceived by both sight and touch. And perhaps we cannot see or feel (i.e. perceive via touch) without seeing or feeling extension and position, i.e. extended things apparently located in various places, but we can perceive, by any of the senses, without perceiving size or shape, solidity, rigidity or impenetrability.

There is, then, a close link between each of the secondary qualities and one particular sense-modality. We perceive, know of, and talk about the secondary qualities we do, only because we

---

[1] It need not be, but it may be. When I look at the match-box in the distorting mirror the shape I see is not the shape of the object. But when I look at the tilted penny the shape I see is the shape of the penny.

have the senses we do. We can conceive of a being who has different senses from us, and who therefore perceives different secondary qualities. He might perceive magnetic attraction, or electrical changes, for example. Or he might be aware of the same physical phenomena as we are, but in a different way. He might, for example, be visually sensitive to sound waves. But we have much greater difficulty in admitting the possibility of a being who did not perceive the primary qualities that we do.

However none of this shows that the secondary qualities are sense-dependent and the primary qualities not. What is sometimes said[1] is that the various secondary qualities are not only proper and tautologous objects of their various senses, but they are also 'internal' or 'cognate' objects of those senses. This terminology is borrowed from the grammar books, where the noun in 'I jump a jump' or 'I hit a hit' is referred to as an 'internal' or 'cognate' accusative. Indeed the apparent similarity between 'I see a colour' and 'I hear a sound' on the one hand, and 'I dance a dance' or 'I play a game' on the other, is taken to show not only that colours and sounds exist only in so far as they are seen and heard,[2] but also that colours and sounds are somehow identical with seeing and hearing![3] Unfortunately this argument, which is a direct refutation of Moore's argument in his *Refutation of Idealism*, involves two important errors.

First, it is a mistake to say that an internal accusative is one where the noun refers to the same thing or activity as the verb. This is certainly true of some internal accusatives; in 'I jump a jump' and 'I hit a hit' both noun and verb refer to the same activity. But there are other examples where this is not so, cf. 'I give a gift' or 'I marry a spouse'. Clearly the gift is not the same thing as the giving, and the spouse is not identical with the marrying. Rather the point is that the thing named by the noun deserves that particular name only because of the occurrence of whatever it is that is described by the verb. The book exists whether I give it or not, but

[1] Cf. Ducasse I.
[2] Ducasse I, p. 229: 'An accusative connate with a given activity exists only in the occurence of that activity'.
[3] Ducasse I, pp. 232–3: 'To sense blue ... is to sense *bluely*, just as to dance the waltz is to dance "waltzily" (i.e. in the manner called 'to waltz'), to jump a leap is to jump "leapily" (i.e. in the manner called "to leap"), etc.'. In other words perception of blue is a kind of perception in the way that a deed of valour is a kind of deed.

C*

it is only in so far as I do give it that it can be called a gift. Similarly the woman exists whether she gets married or not, but it is only in so far as she is married that she can be called a spouse. So even if 'colour' is an internal accusative after 'see', even if 'sound' is an internal accusative after 'hear', it would follow at most that colours can exist only in so far as they are seen and sounds only in so far as they are heard, and not that colours and sounds are, somehow, identical with seeing and hearing.

But second, what justification is there for saying that 'colour' is an internal accusative after 'see', that 'sound' is an internal accusative after 'hear', and so on? To say this is, in effect, simply to beg the question. In fact 'I see a colour' seems not so much like 'I jump a jump' as like 'I eat food'. There can be food which is not eaten, just as there can be colours which are not seen, but to call it food is to think of it as something which can be eaten, just as colours are thought of as something which can be seen. It may be that the secondary qualities have been thought of as internal objects of the respective senses only because people have not distinguished saying this from saying that they are proper and tautologous objects of those senses.

However the fact remains that the secondary qualities are linked to particular senses in a way that the primary qualities are not. Conversely the primary qualities are linked to physical objects in a way that the secondary qualities are not. We can imagine an object with no colour (e.g. a mirror), which emits no sound, or has no smell, but we cannot imagine an object without shape or size. It might be said that we can imagine objects which are not solid, e.g. clouds or puddles, but the question is whether such things are really to count as physical objects. In fact we are reluctant to call them physical objects precisely because they are not solid. This brings out how the primary qualities are linked to physical objects: they are what we might call the 'defining properties' of physical objects. Something will be called a physical object if and only if it possesses the primary qualities, and even if solidity, rigidity, impenetrability are not necessary, extension in three dimensions is.

Why have we chosen these qualities as defining physical objects? One factor might be that primary qualities do not come and go as secondary qualities do. Things may change in shape and size just as they change in colour and taste, but they do not change from having some shape to having no shape in the way that they can change from having some colour to having no colour. But there is a more important point than this. Physical objects are spatio-temporal objects, fundamentally bound up with the spatio-temporal framework in terms of which we individuate, identify and re-identify objects of all types and kinds. And the primary qualities are, in the last analysis, spatio-temporal qualities, i.e. they embody the spatio-temporal character of physical objects. Position involves location in the spatio-temporal framework, extension can be described as location at more than one point in a dimension, shape and size consist in extension in various dimensions, and solidity, rigidity, impenetrability, refer to the occupation of an area of space. And it may be because number is connected with the individuation, identification and re-identification of objects, which works by reference to the spatio-temporal framework, that some philosophers have regarded it as a primary quality. In short, the primary qualities are dimensional properties, and it is this which explains their being the defining properties of, and so essential to, physical objects.

Finally there is the suggestion that the primary qualities are of particular importance to scientists in a way that secondary qualities are not. This is due to several factors, including the fact that the secondary qualities are not essential to the solid bodies which are the physicist's main concern. When the physicist investigates such things as mass, velocity, spatio-temporal co-ordinates and the like he is concerned with the same phenomena we ordinarily describe as solidity, movement, position, and so on. It is easy to see now those who lived around the time of Newton came to regard the primary qualities as more important, and so more real, than the secondary qualities. Even so this suggestion restricts science, in effect, to physics, and even the physicist is, in his own way, interested in sounds, colours, etc.

It might now be said that what the physicist is interested in is not sounds and colours as such, but the physical phenomena, sound- and light-waves, which are the cause of the sounds and colours we perceive, and that this is another difference between primary and secondary qualities. The latter, unlike the former, have an external, physical, cause. In 7.3 I will argue that the physical phenomena are not to be thought of as the causes of the sounds and colours we perceive but rather as, in a definite sense, identical with the sounds and colours we perceive. What we call sounds and colours are, I will suggest, merely these physical phenomena as they appear to the senses of hearing and sight, as they are heard and seen to be. To see a colour is, in a sense (the only sense in which this is possible), to see light-waves, and to hear a sound is to hear sound-waves. And, equally, to see a certain shape or size is to see a certain arrangement of atoms, protons, etc. So the fact that sounds are associated with sound-waves and colours with light-waves does not indicate a genuine difference between primary and secondary qualities. The shape or size or solidity of an object is connected with underlying physical phenomena in just the way that its colour or sound are. Nevertheless the fact remains that the physicist is interested in the physical phenomena which underlie sound and colour rather than in sound and colour as they are perceived to be. There are various reasons for this. Hearing and vision by themselves do not tell us much about the nature of sound and colour, so scientific investigation naturally turns to the physical phenomena as they are independent of these senses. There is also the point that sounds as heard and colours as seen admit only of 'intensive' and not 'extensive' measurement, whereas the physical phenomena that are heard as sounds and seen as colours do admit of extensive measurement. Extensive measurement has definite advantages. Things which can be extensively measured can be directly compared with one another—sizes, shapes, lengths, speeds, etc., can be compared by, so to speak, placing them one against the other—whereas when we are concerned with degrees upon a scale one amount may cancel out or increase another. And extensive measurement is not arbitrary in the way that intensive measurement is. It is an arbitrary matter whether we measure in yards or metres, lunar or calendar months, but when measurements in one system are translated into measurements in another the relationships stay the same. Whether we measure in yards or metres one object will still be twice the length of another, but whether a sound is twice as loud as another, or a colour half as bright, depends on what we pick, arbitrarily, as our standards in terms of which measurement is to be made. We can still translate from one system of measurement into the other, but the relationships between the measured item will vary from system to system, as the relationships between degrees of temperature vary depending on whether we measure in fahrenheit or centigrade.

We have, then, four points of difference between primary and secondary qualities:

(1) The secondary qualities are appearance-determined, while the primary qualities are not.

(2) What secondary qualities we perceive depends on what senses we possess, while the primary qualities are independent of the individual senses.

(3) The primary qualities, as dimensional properties, are the defining properties of physical objects, while the secondary qualities are not.

(4) The primary qualities have extensive magnitude, while the secondary qualities have intensive magnitude.

I do not see that any of these differences warrants our regarding the secondary qualities as sense-dependent, or does anything to suggest an Idealist or Causal theory interpretation of our perception. We must beware of the following mistakes:

(1) The mistake of thinking that because a quality is appearance-determined it exists only in so far as an object appears to someone to have that quality. We might think that because what colour is perceived depends on what colour a thing looks, and since what colour a thing looks can, among other things, depend on the state of the perceiver, it depends on the perceiver that a colour is perceived. But even if *what* colour is perceived does depend, to some extent, on the perceiver, it doesn't follow that it depends on the perceiver *that* a colour is perceived, at least no more than it depends on a perceiver whether anything is perceived. Again the fact that some colours exist only in so far as they are perceived does not mean that all colours do, any more than the fact that some daggers (or items described as daggers) exist only in so far as they are perceived means that all daggers do. Nor does the fact[1] that what colour we ascribe to an object depends on what colour is perceived under certain conditions mean that the object possesses a colour only in so far as it is perceived. The fact that we define what colour an object really is in terms of how it looks may lead us to think that an object cannot be coloured when it is not perceived, for how can it look, e.g. red if it is not being perceived? But to say that an object is red is not to say that it now looks red to anyone. It is only to say that it would look red under the appropriate conditions.

[1] Cf. 6.2.

(2) The mistake of passing from 'What is the real colour?' to 'Is it really coloured?', i.e. of passing from the fact that things can and do appear to have different colours in different circumstances, and the problem of establishing which of these colours is the real colour, to the quite unwarranted conclusion that the object has no real colour. A similar mistake might be made with primary qualities.

(3) The mistake of thinking that the secondary qualities are 'internal' objects after the verbs of perception, and so thinking that they exist only in so far as there is perception, not to mention the mistake of identifying these qualities with types or species of perception itself.

(4) The mistake of thinking that since we perceive, conceive and know of the secondary qualities we do because we have the senses we do, it is only because we have these senses that these qualities exist. This is the mistake of thinking that because knowledge of $x$ depends on $y$, $x$'s existence must also depend on $y$.

(5) The mistake of thinking that because the secondary qualities are not essential properties of physical objects they are not real properties of physical objects.

(6) The mistake of thinking that because physicists are not particularly interested in the secondary qualities as such, the secondary qualities do not belong to the physical world in the way that the physical phenomena, in which the physicist is interested, do.

(7) The mistake of thinking that the secondary qualities are caused by those physical phenomena.

There is more to be said about 1, 6 and 7. We shall be returning to these topics in the next two chapters.

## 5.2 THE ARGUMENT FROM SENSATIONS

The argument from sensations is the argument that the things we perceive are sensations, and as such exist only when they are perceived. Clearly the crucial question is whether we do perceive sensations, in any sense from which it follows that they exist only when perceived. The main source of error here has been the multiple ambiguity of the term 'sensation'.

'Sensation' has two ordinary uses.[1] First, we use it to refer to one particular kind of perception, perception via the tactual and

[1] Cf. Ryle III.

kinaesthetic senses, as when we ask someone whose arm has been anaesthetized whether sensation has returned to it yet, i.e. whether he can feel it, or with it, yet. Here 'sensation' is like 'vision' or 'hearing', and we do not talk about 'a sensation' or 'sensations' in the plural at all. The second non-technical use is where we refer to identifiable and locatable states and affections of the body, such as pains, tickles, itches, feelings of heat and cold and so on, as 'sensations' (notice the plural). To avoid confusion we might call these 'bodily sensations'.

Philosophers have developed, and often failed to distinguish, two further uses of the term 'sensation'. Sometimes we want to draw a distinction between perception, in the full sense of the word, and a basic process of awareness, independent of such things as recognition, judgement or interpretation. This process is often referred to by means of the verb 'to sense', and this leads easily to talk of a process of 'sensation', i.e. sensing (what I call 'sensory awareness'). 'Sensation', in this sense of sensing, would seldom, if ever, be used in the plural. But it is just a short step from using 'sensation' as the name of this process, to using 'sensation' as the name for what is sensed, and in this latter use the noun will often take the plural. It is in this sense of 'sensation' that we speak of visual, auditory or olfactory sensations. I take it that 'sensation' here comes to mean much the same as 'sense datum', although we shall see (11.3) that it is a mistake to think of the sense datum as what is sensed as opposed to what is perceived, for the simple reason that what is sensed and what is perceived are normally one and the same thing.

In discussing the various factors which might mislead us into taking it for granted that the things we perceive are sense-dependent, I mentioned the tendency to slip from talk about what we perceive to talk about perception itself. Although the two are obviously different, philosophers do tend to assume that what is true of perception—that it is, in some sense, mental, private and sense-dependent—must also be true of what is perceived. Some have even been led to think that what we perceive is, somehow, simply an aspect of the perception.[1] The oddity of this is usually concealed by the use of such words as 'perception' and 'sensation'.

---

[1] This mistake is still made by those who speak about 'adverbial' analyses of perception, as though what we perceived could be just a way of perceiving, cf. Broad III, Hirst I.

We slip from talk about our perception or sensation, to talk about our perception*s* and sensation*s*, where as the plural indicates we are not talking about the process of perceiving or sensing, but about what we are aware of (cf. the similar shift from 'utterance' in the sense of the act of uttering, to 'utterance' in the sense of what is uttered).

We will see that this error, this failure to distinguish perception from what is perceived, is more common than one might have supposed possible. It was first pointed out by Moore in his *Refutation of Idealism*, although this major point of his article is easily obscured by its difficult terminology and the confusing use of terms like 'idea' and 'sensation'. The argument is clearly stated by Stace:[1] 'If we compare a green sense datum with a blue sense datum, we find a common element, namely awareness. The awareness must be different from the green because awareness also exists in the case of awareness of blue, and *that* awareness, at any rate, is not green. Therefore, since green is not the same thing as awareness of green, green might exist wthout awareness'. In *Some Main Problems of Philosophy* Moore added two further points: (1) It is at least logically possible that the colour I see should continue to exist after I see it, but it is not logically possible for my seeing to continue to exist when I stop seeing; (2) It is logically possible that the colour I see belongs to an external object, but it is not logically possible that my seeing belongs to an external object.

These arguments do not show that what is perceived does exist independently of being perceived. Moore does not so much refute Idealism, as uncover one mistake which might tempt us to think that, so far as the things we perceive are concerned, to be is to be perceived. What the arguments do show is that what is perceived must be distinguished from the perception of it. This means that we must guard against the confusion between 'sensation' in the sense of 'sensing', and 'sensation' in the sense of 'sense datum'.[2]

---

[1] I, p. 369, cf. Moore IV, pp. 30–1, p. 44; and Russell I, pp. 41–2. Moore's argument has been challenged by Ducasse (1). See p. 73 above.

[2] Hamlyn I is one example of this confusion. Although the word appears in the very title of his book, Hamlyn never explains what he means by 'sensation'. Its coupling with 'perception' suggests that he means 'sensing', but he talks interchangeably (cf. p. 196) about 'sensation' and 'sensations'. The same thing is true of Ryle's distinction between observation and sensation (I, ch. 7), although Ryle admits to misgivings about his use of the 'official' sensation terminology.

There is a difference between saying that all perception involves sensation, where that noun does not usually take a plural, and saying that all perception involves having sensations, even though both may be true. If we do not distinguish 'All perception involves sensation (sensing)' from 'All perception involves sensations (sense data)', we may find ourselves led from the fact that sensing is, in some sense, private, mental and sense-dependent, to the conclusion that sense data are private, mental and sense-dependent, and hence to the conclusion that sense data are percepts. It was to avoid this sort of confusion that I preferred to speak of 'sensory awareness' rather than of 'sensing' or 'sensation'.

Another error we must guard against is the assimilation of visual, auditory and other 'perceptual' sensations to bodily sensations. Their unexplained use of 'sensation' makes the theories of Berkeley and Mill, for example, more plausible than they would otherwise be. If, as is usually held, bodily sensations exist only in so far as they are perceived, then it will seem that to say we always perceive sensations will be to say that we always perceive sense-dependent entities. But clearly perception does not always involve sensations in the sense of bodily sensations. Seeing, for example, does not involve having sensations in the eyes, or anywhere else. If we do not distinguish bodily sensations from perceptual sensations we might take it for granted that perceptual sensations exist only when perceived. And once we see the distinction, any attempt to argue that perceptual sensations are bodily sensations will be grossly implausible.

Berkeley[1] attempts to argue that perceptual sensations are bodily sensations, but much of what he says is scarcely worth discussing. He says that tastes and smells are pleasant and unpleasant, pleasing and displeasing, as if it followed from this that they are sensations. Even apart from the question of whether all tastes and smells are pleasant or unpleasant, and the question of whether to say that a taste is pleasant or a smell unpleasant is to say that they are identical with pleasantness or unpleasantness, there is the point that pleasantness or pleasure are not bodily sensations in the first place. The one argument that seems to have any strength is

---

The contrast with 'observation' suggests that by 'sensation' he means 'sensing', but he too talks equally about 'having sensations'. We shall see (11.4) that this confusion invalidates Ryle's argument against sense data theories.

[1] II, First Dialogue.

the argument about heat and cold, which is not surprising when we remember that heat and cold are sometimes perceived as bodily sensations felt and located within the perceiver's own body. But they can also be perceived as sensible qualities belonging to external objects; it is possible to feel the heat of an object without having a sensation of heat in one's own body. Someone who has been in the cold may feel the heat of the fire with his hands, without feeling heat in his hands, even with his hands continuing to feel cold. But Berkeley's argument is that not just some but all heat and cold that we feel are bodily sensations. He makes two points.

First, he says that to feel intense heat or cold is to feel a pain, and pains are, of course, bodily sensations. We might want to say that heat or cold is the cause of the pain, and not identical with it. Berkeley's answer is that we feel both heat and pain, and yet are aware of but one simple uniform thing, not two distinct things, which means that the two must be identical. This argument cannot work. The fact is that we all know and can feel the difference between heat and pain. So if we put our hand in the fire we will feel what we know to be heat, or we will feel what we know to be pain, or we will feel distinct sensations of both at once, or, most probably, we will feel a single sensation compounded of heat and pain. Yet the fact that heat and pain might combine to form a distinct sensation, different from the individual sensations of heat and pain, does nothing to show that the two are, in the end, identical. Similarly the fact that a trumpet and a trombone might combine to form a distinct sound, different from the individual sounds of trumpet and trombone, does nothing to show that the sounds are, in the end, identical.

Second, Berkeley, asks why we locate heat in the fire when we do not locate pain in the knife. There seem to be three important reasons for this. First, the pain is not something I feel just because the knife happens to be near or even touching me. I feel pain only if something is done with the knife—and a different thing will have to be done with the knife for every person who is to feel pain—while all that is necessary for me to feel the heat of the fire is that I be near it. Second I may well continue to feel the pain even though the dagger is thousands of miles away, or even destroyed, so, naturally, we do not ascribe the pain to the dagger. To this it may be said that I may well continue to feel warmth even though

I have gone away from the fire, or even if it has gone out, but then, significantly enough, the heat I feel is said not to be the heat of the fire but a bodily sensation located in my own body. The heat ascribed to the fire, as opposed to heat transferred to my body or the air by the fire, cannot be felt if we go away from the fire, or if the fire goes out. Third, the intensity of the heat varies as we move closer to or further away from the fire, whereas the intensity of the pain has nothing to do with the nearness, nature or even existence of the knife.

Professor White has pointed out to me that a pain in my eyes caused by a powerful searchlight beam may be like the heat of the fire in all these respects, and yet we still ascribe the pain to ourselves as a bodily sensation, and do not locate it in the searchlight beam. One reply might be that the pain-of-the-searchlight example is still not exactly like the heat-of-the-fire example because the pain might continue, if only for a short time, after the light has been turned off, whereas the heat of the fire is, necessarily, not felt once we go away from the fire. If we continue to feel warmth then, *ipso facto*, it cannot be the warmth of the fire that we then feel. This fact, together with the phenomenological similarity between what we feel in the knife case and what we feel in the searchlight case, might explain why we say merely that the searchlight beam is painful, and not that it possesses pain in the way that the fire possesses heat. But even this is not the full story. The thing about the fire is that its warmth can be communicated to something else, a stone perhaps, and we can feel the warmth of the stone without going near the fire. If the searchlight made the things it illuminated painful, so that we could feel pain when we saw those things even after the light had been turned off, then we might think of pain as a quality of the searchlight beam.

### 5.3 BODILY SENSATIONS

At this point we might leave for a moment the argument between the theories of perception, and say something about the status of bodily sensations as objects of perception. Despite their name bodily sensations are often thought of as having rather a ghostly air, as being mental or quasi-mental phenomena. This, of course, supports the argument from sensations. If it is agreed that the bodily sensations we perceive are some sort of mental entity it may be easier to accept that other things we perceive are mental entities too. So in this section I want to consider how bodily sensations differ from other perceptual objects.

The first suggested difference might be that bodily sensations are not perceptual objects at all, that unlike tables and chairs, sounds and smells, shapes and colours, bodily sensations cannot be perceived. Certainly we do not normally talk about perceiving bodily sensations, but for that matter we don't normally talk about perceiving sounds and smells either. 'Perceive' is ordinarily used, if at all, of vision. Nevertheless 'perceive' is used by philosophers to cover all the various modes of sensory awareness so I see no reason why we should not talk about perceiving bodily sensations; feeling a pain is as like hearing a sound as hearing a sound is like tasting a taste.

However some may object to talk about perceiving bodily sensations on the grounds that bodily sensations are not *objects* of perception so much as *types* of perception. The word 'sensation', in particular the confusion between 'sensation' in the sense of 'sensing' and 'sensation' in the sense of 'bodily sensation', is largely to blame here. There is also the fact that feeling a bodily sensation involves feeling something in one's body, which may lead us to equate what goes on in us with the perception, to equate what I feel in my toe when I kick a stone, with my perception of the stone. But what goes on in my toe is not perception; what goes on in my toe has to be perceived, felt, as much as the stone does. It may also be necessary to point out that, despite talk about 'adverbial' analyses of perception, to say that something exists only in so far as it is perceived, which is something most philosophers have wanted to way about bodily sensations, is not to say that it is a type or way of perceiving. Moore's arguments in the *Refutation of Idealism* and elsewhere apply to our awareness of bodily sensations as much as they do to our awareness of colours.

So if the pain that I feel is to be compared with anything in the case of, e.g. seeing a tree, it is with the tree and not the seeing. Both the tree and the pain are given a location, are perceived as being in a certain place, and we talk of being conscious of, and even of paying attention to and coming to notice, the pain just as we talk of being conscious of, paying attention to and coming to notice the tree. The pain is what I feel just as the tree is what I see, and I do not notice the awareness of the pain any more than

I notice the awareness of the tree. Bodily sensations, then, are a kind of perceptual object, and our question remains: what differences are there between them and other perceptual objects?

A second suggested difference might be that bodily sensations, unlike other perceptual objects, are incorporeal or ghostly. This, together with their privacy, is our reason for regarding bodily sensations as some form of mental phenomenon. Now when we say that mental phenomena, like thoughts or beliefs or thinking or perceiving, are incorporeal, we might mean any or all of three things: that they cannot be perceived; that they have no physical location; or that they are not solid. Of these only the third is true of bodily sensations, but this does not indicate any genuine difference between bodily sensations and other perceptual objects. Although I do not bump into or fall over pains or itches, the same is also true of shadows, sounds and smells. It may be significant that philosophers have often, perhaps for this very reason, tended to regard sounds and smells and the like as less real than physical objects.

Next, bodily sensations seem to differ from other perceptual objects in that they are private in the rather special sense of being perceptible by one person alone, and in that they are perceived and located within the bodies of animate beings only, the bodies of the beings who feel those sensations. Certainly this is a difference between bodily sensations and other perceptual objects, but is it a necessary difference? It is, presumably, true by definition that bodily sensations, qua bodily sensations, are felt and located in animate bodies, etc., but is it necessarily true that such things as pains and kinaesthetic sensations are bodily sensations in this sense? Could pains, kinaesthetic sensations, etc., be felt and located in inanimate objects? And could pains, kinaesthetic sensations, etc., be public, perceptible by more than one?

The answer is that they could. We feel pains, etc., only within our own bodies because of the nature of the sensory mechanism involved, because the network of nerve fibres by means of which we feel pains does not extend beyond our own bodies. I have argued elsewhere[1] that if we could link our own nervous systems with those of other people we could feel the pains, etc., that they feel, and feel them in their bodies. Similarly if inanimate objects contained the appropriate nerve fibres it might be possible for us to

[1] III.

link our nervous systems with those nerve fibres and so feel pains, etc., in those inanimate objects. This would not mean that those objects were animate after all, so long as they did not themselves feel the sensations we feel in them. So it is not a necessary fact that pains, kinaesthetic sensations and the like are private and felt and located in animate bodies alone; this is but a consequence of our nervous system's being as it is. The view that bodily sensations are necessarily private is due, I think, to a confusion between different types of privacy, between saying that something is necessarily private in the sense that it (logically) could not belong to someone else ('l-privacy'), and saying that it is private in the sense of perceptible by one person alone ('m-privacy'). Pains are private in the logical sense, but this does not mean they are necessarily m-private.

The next suggestion is that we cannot be mistaken about the bodily sensations we perceive, whereas we can be mistaken about other perceptual objects, i.e. can think they are other than they are. But surely it is possible to be mistaken about our bodily sensations? Take the initiation trick where the unfortunate subject is blindfolded and told that he will be branded with the red-hot poker he can hear, and smell, being heated for the purpose. A piece of ice is pushed into his bare stomach and, naturally enough, he screams. There seems a reasonable case for saying he mistakes a sensation of cold for one of warmth. To be sure he will probably realize the deception almost immediately, but if he dies of fright on the spot it seems he will die thinking that he feels a sensation of intense heat, or pain, when in fact he does not. Or suppose that I have been suffering from a severe, but blessedly intermittent, toothache. I get a tingling in my cheek and for a moment, or even longer if my attention is elsewhere, I think it is the toothache again. And cannot I mistake toothache for headache? Or suppose 'somebody has experienced severe pain in the region of his heart. On describing the nature of the pain to a doctor, he learns that it is the symptom of a serious heart condition. On another occasion he gets a *different* sort of pain in the same place. Is it not conceivable that, on getting this new sort of pain, he may panic, so that at first it *feels to him* as if it is the old pain again? Again, it is not possible, especially if we panic, for a pain to feel to us to be more intense than it really is?'.[1] In fact children often make just this

---

[1] Armstrong II, p. 55.

mistake; the sight of blood makes them think the grazed knee hurts more than it does. Even adults have been known to say 'For a moment I thought he had really hurt me'. Finally there seems nothing wrong with the suggestion that a person might make mistakes about his kinaesthetic sensations. I might think that these are the sensations I get when my hands are forced behind my back and in a figure 3 movement, when in fact they are the sensations I get when it is forced in a figure 8.

Two factors make it easy to think we cannot be mistaken about our bodily sensations. First, since we alone can perceive our bodily sensations it is difficult and uncommon, though not impossible, for others to correct us about them. I cannot correct your judgement about your pain, but the doctor may say to the man with the heart condition 'Surely it can't be the same pain again?' and, reassured, the patient may realize that he is right. And second, bodily sensations are, like colours, appearance-determined, which means that the way to tell what a bodily sensation is like is, simply, to feel it, and notice how it feels, how it appears. But just as we can be mistaken about what colour a thing appears to be (cf. 11.5) so we can be mistaken about how our bodily sensations appear.

The claim that we cannot be mistaken about bodily sensations is linked with the claim that we cannot be mistaken about their existence. I think it is possible for a person to think he feels a pain when he feels nothing (e.g. a frightened child who sees blood on his knee), just as it is possible for a person to think he sees when he sees nothing. What is true is that, ordinarily, we do not draw a distinction between real sensations and hallucinatory sensations, although it is not difficult to think of cases where we might want to draw this distinction. We might, for example, speak of 'phantom' sensations or psycho-somatic pains as hallucinatory.

However although we may not speak of sensations which are perceived but do not really exist, I think we can and do speak of sensations which exist without being perceived. It is usually held that the notion of an unperceived sensation is a contradiction in terms. Two things account for this mistaken dogma. First there is an unfortunate tendency to take pains as the paradigm example of a bodily sensation. Pains are, by definition, obtrusive. I might tickle someone without his knowing it but I cannot hurt him without his knowing it, because 'hurt', unlike 'tickle', requires, logically, that the recipient notice it. No matter what I do to you if it isn't drastic enough to secure your attention it isn't drastic enough to count as causing you pain. The only cases where we want to talk of an unperceived pain, are significantly, cases where our attention is diverted. If a man gets badly burnt while helping at a fire he may not notice the pain until the fuss and the flames have died down, but there is a case for saying his hand was hurting all the time. Even so ordinary language is ambiguous on this point: 'While you were talking I didn't notice the pain' suggests that the pain continued unnoticed, but at the same time seems equivalent to 'While you were talking the pain stopped'. But even if we are to reject the notion of an unfelt pain this is not to reject the notion of an unfelt sensation. Cannot there be unfelt itches and tickles? Suppose I find myself scratching my leg and realize that it is itching; is it not tempting, indeed natural, to say that there was an itch there even before I noticed it, that I scratched it because it itched even though I had not yet felt that itch? Or is there anything odd in talk about the various kinaesthetic sensations which must have occurred in my fingers and wrists as I typed the last page, even though I did not perceive a single one of them? And do I always perceive the tactual sensations I get whenever I touch something? These examples help counter-balance the usual emphasis on pains, and suggest that there is, after all, nothing wrong with the notion of an unperceived sensation. An unperceived sensation, like an unperceived sea-shell, is something which could have been perceived, but for various reasons was not.

The second factor which leads people to reject the notion of an unperceived sensation is that old confusion between perception and what is perceived. We talk, confusingly, about a feeling of strain in a muscle, but the strain is what we feel and not the feeling, the awareness, of it. So the fact that the awareness lasts only so long as we feel the strain does not mean that the strain lasts only so long as we feel it.

But although we can talk about sensations existing unperceived the fact remains that we do not draw a full distinction between real and merely hallucinatory sensations. We could draw such a distinction but we do not. Why? Once again there seem to be two reasons. First the fact that bodily sensations can be felt only by the person to whom they belong makes the suggestion that he might feel a sensation which does not exist, or that he might have

THE ARGUMENTS TO SENSE-DEPENDENCE    89

a sensation which he does not feel, sound impudent to say the least. But once something is known about the physiological mechanism by means of which these sensations are produced and felt, a way is provided for saying that a person feels a sensation which doesn't exist, or saying that there is a sensation in his body which he does not feel. Nevertheless our language and our concepts preceded this physiological information, so they do not allow for a distinction—which physiologists, psychologists and philosophers might well want to make use of—between real and merely hallucinatory sensations.

The second reason is that whatever a physiologist may want to mean by a bodily sensation, what *we* mean by a pain or the stretching sensation I feel in my elbow as I straighten my arm is not the stimulation of nerve ends or anything like that, but something, *sui generis* if you like in the way that colours are *sui generis*, of the sort we perceive normally, but not always or solely, when those nerve ends are stimulated. We may even want to say that what we feel is something which is produced by the stimulation of those nerve ends, but this way of talking is misleading. It immediately suggests that sensations are things produced 'in the mind', which raises several difficulties. There is also the point that 'produce' naturally suggests some form of physical causation when no physical link between the stimulation of the nerve ends and the sensation can be found, if only for the reason that the physiologist cannot find the sensation in the way he can find the stimulation of the nerve ends. I would prefer to say not that the stimulation 'produces' the sensation but that the sensation is, *nothing but* the stimulation of the nerve ends, as it appears to the perceiver via the relevant bodily sense. Of course 'pain' doesn't *mean* 'stimulation of such and such nerves', for the discovery that pains are, are 'produced' by, stimulation of nerve ends was a scientific, factual, discovery. However, if we are to say that pains are, in effect, the way in which certain physiological phenomena appear to us—together with other phenomenologically similar things such as psycho-somatic pains and pains in phantom limbs—we are virtually saying that pains are appearances, although not necessarily 'mere appearances' in the pejorative sense. And since there is something very strange about the idea of an appearance which appears to no one we find ourselves inclined to say that the pain, as opposed to the physiological phenomena which 'produce' it (i.e. present this appearance) has to appear to someone, i.e. exists only when felt.

But now consider the parallel case of sounds. Whatever the physicist may want to mean by sound, what we mean is not a certain sort of wave motion but something, *sui generis* if you like in the way that colours are *sui generis*, of the sort we perceive normally, but not always or solely, when our ear-drums, etc., are stimulated by such wave motion. We may even want to say that what we hear is something which is produced by the action of this wave motion on our ears, but this way of talking is misleading. With sounds, presumably because they are not private in the way that sensations are, we are not so tempted to talk of them as produced 'in the mind', but nevertheless 'produce' naturally suggests some form of physical causation, even though no physical link between the wave motion and the sound can be found, if only for the reason that the physicist cannot find the sound in the way that he can find the wave motion. I would prefer to say not that the wave motion 'produces' the sound but that the sound is *nothing but* a certain sort of wave motion, as it appears to the perceiver via the sense of hearing. Of course 'sound' doesn't *mean* 'such and such a wave motion', for the discovery that sounds are, are 'produced' by, this wave motion was a scientific, factual, discovery. Now if we are to say that sounds are, in effect, the way in which certain physical phenomena appear to us—together with other phenomenologically similar things, hallucinatory sounds—we are virtually saying that sounds are appearances, although not necessarily 'mere appearances' in the pejorative sense. And since there is something very strange about the idea of an appearance which appears to no one we find ourselves puzzled by such questions as 'When a tree on a deserted island falls down, does it make a sound?'. Interestingly enough people are genuinely bemused by this question when they would be simply intolerant of the suggestion that the radio stops making a noise just because everyone goes out of the room.

Nevertheless in so far as we are prepared to talk about sounds existing unperceived—and of course we are—we should equally be prepared to talk about sensations existing unperceived. The difference between sensations and sounds is only that sensations are always associated with, and felt in, human bodies, so that whenever conditions are such that a sensation might be felt, there is always someone around to feel it.

To sum up: the crucial difference between bodily sensations and other perceptual objects is that they are as a matter of fact, though not necessarily, perceived and located in animate bodies only, and perceived only by the being whose body they are located in. These facts explain our inclination to insist, as a conceptual point, that bodily sensations exist whenever and only in so far as they are perceived, i.e. our tendency not to draw a distinction between real and perceived existence as regards bodily sensations. It also explains why we regard bodily sensations as appearance-determined.

## CHAPTER 6
# THE ARGUMENT FROM ILLUSION

### 6.1 NON-VERIDICAL PERCEPTION

Idealism, and to a lesser extent the Causal theory, often appeals for support to the facts of what philosophers refer to as 'appearance' and 'illusion'. The suggestion is that the occurrence and nature of non-veridical perception shows that we perceive nothing but percepts. It also shows us that if seeing is believing it is not necessarily knowing, and so it gives rise to such questions as how we can know what what we perceive is really like, how we can know whether what we perceive really exists, how we can know what things are like independent of our perception and so on.

Discussion of these issues has been confused by a failure to distinguish the different ways in which perception can 'go wrong'. So let us begin by distinguishing four different types of non-veridical perception:

(1) I sense something that does not really exist (as with Macbeth's dagger).

(2) I sense something that does really exist, but I take it to be something else, something else that does not really exist (as when I take a vine to be a snake).

(3) I sense something that does really exist but, objectively, it is different from what I sense it as being (as when I see a red tomato but, through colour-blindness, sense it as grey).

(4) I sense something that does really exist, and I sense it as it is objectively, but I take it to be what it is not (as when, in the Muller-Lyer illusion, I see two lines which are in fact equal, but take one to be longer than the other).

There are some comments to be made about these four cases.

(1) Two points need to be made about this case. First, as I argued in 1.2, this is something which can and does happen. We must beware of the temptation to define the word 'perceive' in such a way that it becomes logically impossible to perceive what doesn't exist, and thus avoid having to discuss hallucinations. We can now see that this move rests, in part, on a failure to distinguish between the first two types of non-veridical perception. It would

be a misuse of the ordinary verb 'see' to say of the man who takes vine to be a snake that he saw a snake (we would rather say that he 'saw', i.e. thought he saw, a snake), but it is not a misuse to say that Macbeth saw something, something which he described as a dagger.

The second point is that it follows from my definition of 'really exists', i.e. 'exists independently of our perception of it', and from my account of non-veridical perception, that perception of afterimages and buzzings in the ears is non-veridical perception. This is rather odd. It would be preferable to put these things to one side, to introduce a third category of, say, 'a-veridical perception', perception where the veridical—non-veridical distinction does not apply. But it is rather difficult to provide an adequate definition of 'a-veridical perception'. We might, for example, say that perception is a-veridical where the item perceived is of a type where items of that type are never veridically perceived. But apart from the point that whether perception counts as a-veridical or non-veridical will then depend on how we describe what we perceive (whether I describe it as an after-image or as a circular brown patch), there is the difficulty that hallucinations are never objects of veridical perception, which means, on this account, that perception of hallucinations is a-veridical perception! I think the answer, if not the solution, to the difficulty is simply that there is no strict boundary between non-veridical and 'a-veridical' perception. And since it is the distinction between veridical and non-veridical perception that we are concerned with let us ignore these complications.

(2) Here, as in 1, what I would describe myself as perceiving does not really exist, but unlike 1 I do not perceive any such thing as I describe myself as perceiving. I have argued that it is here that the scare-quotes use of the perceptual verbs really belongs, for I am not acquainted with a snake, i.e. do not sense it, in the way that Macbeth was acquainted with his dagger. But this case of thinking that one perceives an *x* must be distinguished from the case of thinking that one perceives an *x* when in fact one perceives nothing. When I 'see' a snake there is something that I see, a vine, and it is because some perception is involved that it is correct to talk about perception, albeit non-veridical perception, at all. But the lunatic who thinks he sees Napoleon's army advancing behind him is not seeing something which he mistakes for Napoleon's

army, he is seeing nothing of the kind at all. Unlike my 'perception' of the snake his 'perception' is no more perception than a putative judge is a judge.

(3) There is an important difference between my perceiving a red tomato as grey because I am colour blind, and my perceiving a grey wall as red because of the light shining upon it. In neither case is what I perceive as I perceive it to be, but only in the first case does my perception count as non-veridical. If, under the red light, I saw the real colour of the wall, grey, then I think we would say that my perception was *non*-veridical! This is the point of the adverb 'objectively'. An object is objectively red if, under the existing conditions of observation, it is perceived as red by everyone who does not suffer from some special idiosyncracy such as colour blindness. And, as in our example, an object may be objectively red without being really red. We will come back to the question of what it is for an object to be really $x$. The point for the moment is that the fact that the quality I perceive is not the real quality of the object does not mean that my perception is non-veridical.

(4) Not all 'misperception-that', mistakes of judgement about what is perceived, count as non-veridical perception. We saw in 2.3 that perception-that can go beyond what is perceived, and in such cases mistake does not involve non-veridical perception. If I perceive a vine and take it to be a snake we may say, ironically, that I 'saw' a snake, but if I perceive a sketch by Braque and take it to be a sketch by Picasso I don't think anyone would want to say that I 'saw' a sketch by Picasso, or that my perception was non-veridical.

Three final points: First, the differences between these four types of non-veridical perception may not be as clear-cut as I suggest. It may often be difficult to tell which type a particular instance of non-veridical perception belongs to. Second, we might say, using the distinction drawn in 2.3, that types 1 and 3 involve non-veridical perception in the weak sense of 'perceive', and 2 and 4 involve non-veridical perception in the strong sense. And finally, it is not perception as such which is veridical or non-veridical, but perception of particular things. When I see the scarecrow in the fog my perception is veridical in so far as I perceive a vague looming shape, and non-veridical in so far as I take it to be a man. Similarly, Macbeth's perception of the dagger was non-veridical perception, but his perception of his hand was veridical perception.

## 6.2 REALITY

Philosophers have often wanted to distinguish between Reality and Appearance or Illusion. These last two labels can be very misleading. The basic distinction they want to make, I think, is that which I drew in 1.2 between 'real existence' and 'perceived existence'. Reality consists of whatever exists independently of our perception; it has to be distinguished from that which exists only in our perception of it.

This traditional notion of Reality has been sharply criticized by Austin.[1] It seems to me that his criticism relies mainly on a failure to distinguish between two different sense of 'real'. We can distinguish between something which is not a real dagger because it doesn't really exist, and something which is not a real dagger because it is made of plastic. There is a difference between 'real' in the sense of 'really existing', and 'real' in the sense of 'genuine' (as opposed to 'counterfeit', 'imitation', 'pretence', etc.). When philosophers talk about Reality it is the former sense of 'real' that they have in mind, but Austin's discussion deals only with the latter sense. He says, for example, that 'real' is 'substantive hungry', in that the question 'Is it real or not?' must be backed by an answer 'A real *what*?' if it is to be answered or even understood (and the answer to 'A real what?' determines what we are asking when we ask 'Is it real or not?'). But when we ask of Macbeth's dagger whether it was real, the question 'A real what?' is quite irrelevent. It makes no difference whether it is a dagger or a paperknife or a scalpel; the question is whether it, whatever it is, really exists. In this sense 'real' is not 'substantive hungry' at all.

It might be argued that there is a close, indeed necessary, connection between these two senses of 'real'. After all physical objects such as tables and chairs are our paradigm examples of real objects, i.e. external objects, and the argument goes, anything which does not really exist is not a real physical object, in the other sense of 'real'. That is, a dagger which, like Macbeth's, doesn't really exist cannot be a real, genuine dagger. At best it is an imitation or counterfeit dagger; something which looks like, but is not, a dagger, just as a decoy duck looks like, but is not, a

[1] II, Ch. 8.

duck. The question is, in effect, whether it is true by definition that physical objects are external objects, and this question is worth considering in itself. We have seen that there can be external objects which are not physical objects—sounds or smells for example—but what of the reverse? Are hallucinatory physical objects a kind of physical object, or are they no more physical objects than a putative judge is a judge, or a blown-up bridge a bridge?

On the one hand it seems natural to say that Macbeth knew that what he saw was a dagger and not, say, a breadknife. His question, no doubt rhetorical, was not what it was but whether it really existed. And if he *knew* that it was a dagger it must be a dagger. But, on the other hand, daggers are physical objects, and physical objects can be touched as well as seen, while this 'dagger' could not be touched. How could it be a dagger when it was neither sharp nor blunt, could not be used to stab or cut, had none of those characteristics which distinguish daggers? It begins to look as though Macbeth would be wrong to call it a dagger, as though a real (genuine) physical object must be a real (really existing) physical object.

Even so this is not true of all objects of perception. It might be said that it is difficult to imagine what a non-genuine, imitation, sound might be like, and that this supports the view that there is something wrong with my radical distinction between two senses of 'real'. But suppose we think, instead, of individual types of sound, or individual colours. Surely we can have a real (genuine) red that does not really exist, e.g. the red of an after-image, or a real (genuine) ringing that does not really exist, e.g. the ringing in my ears? 'Real', in the sense of genuine, does not always entail 'real' in the sense of really existing.

However I am not convinced that 'real (genuine)' entails 'real (really existing)' even for physical objects, although, assuming the truth of Realism, 'All real (genuine) physical objects really exist' might be a true empirical generalization. Even given Realism I think there could be, highly unusual, circumstances in which we might well talk of real, genuine, physical objects that do not really exist. The main reason for insisting that 'Physical objects are external objects' is true by definition is that physical objects can, by definition, be touched as well as seen, and it is assumed that anything that can be touched, or more accurately anything that is solid, must really exist. But suppose, as may happen, that a man has tactual hallucinations to fit his visual ones. Perhaps, despite his hallucinations, he can still be made to pass through the area where he feels and sees an object. But suppose that when we try to push him through this place we find, to our consternation, that although we feel and see nothing there, and can ourselves pass through the spot without let or hindrance, he just will not go, that his body flattens out as if pushed into something solid and, we might add for good measure, is marked with abrasions. Fortunately for our conception of the external world such strange things do not happen, but they are logically possible, and in so far as there is a case here for saying that the man perceives a real, genuine (tangible and solid), physical object which does not really exist—and I think there is a case for saying this—so far it is logically possible for there to be a real, genuine, physical object which does not really exist, i.e. so far 'physical object' does not entail 'external object'. Dream objects may be another case in point: if in my dream I see and feel and grasp and stab with a dagger, can it not be said that the dagger was a real, genuine, dagger, although, of course, not one that really existed?

However even if we agree that 'physical object' does not entail 'external object' the fact remains that Macbeth's dagger, for one, was not a physical object and, therefore, not a dagger, because it could not be touched, was neither sharp nor blunt, etc. This is rather surprising, because we still want to say that what Macbeth saw was, in some sense, a dagger, and not something else. I think it helps to avoid confusion if we follow White[1] in distinguishing an *identification* of what is perceived from a *description* of what is perceived. 'Describe' normally means, in a wide and general sense, something like 'say something about', but I think something is to be gained by narrowing its sense and contrasting it with 'identify' (in the sense of 'identify as' rather than the sense of 'identify with'). Roughly speaking to identify something is to say what it is, to assign it to some class or category, to fit it with a label or name; and to describe it is to say what it is like. Saying that a thing has certain properties, or that it is like various things, is not saying what it is, although it may follow from the fact that it has certain properties that it is a thing of a certain sort.

So we might say although it would be a mistake for Macbeth to

[1] I, pp. 176 ff.

identify what he sees as a dagger, it is not a mistake to describe it as one. What he sees is not, for reasons already considered, a dagger, but it is like a dagger, and so 'A dagger' is a perfectly accurate description of it. Similarly it is quite legitimate for the man looking at a picture in a book or on a cinema screen to describe what he sees as a horse, even though no such thing can be felt, let alone ridden, fed or shod. Even so I am a little unhappy about calling these 'descriptions', even in White's sense. 'It is a dagger' looks more like an identification than a description. If to describe is to say what the thing is like perhaps we should rather say 'It is like a dagger', just as we could also describe it as 'like a breadknife', on the grounds that daggers are like breadknives. Indeed why do we accept 'It is a dagger' and not 'It is a breadknife' when what Macbeth saw was (identification) neither but was like (description) both? The answer seems to be that within a certain range or context, i.e. speaking in terms of hallucinations, we can identify what Macbeth saw as a dagger, just as, speaking in terms of pictures, we can identify what the man in the cinema sees as a horse. These are what we might call 'context-bound' identifications. 'It is a dagger' would be false if it were a straight-forward identification, but as a context-bound identification, an identification of the sort of hallucination Macbeth saw, it is correct. I think it is misleading to call 'It is a dagger' a description, for it seems undeniably to say what what is seen is, and not merely what it is like.

So far we have been concerned with 'real' as applied to objects, but equally important is the usage where it is applied to qualities or properties. When we say of something that, in the sense which is of special interest to philosophers, it is really red, or loud, or heavy, what we mean is that it *is* red, loud or heavy, as opposed to merely *appearing* red, loud or heavy. The qualities we perceive when we perceive a thing, and the qualities which that thing appears to have, may be very different from the qualities which the object actually possesses, the qualities we ascribe to the object. A thing which is white may look red; a thing which is round may look elliptical; a sound which is loud may sound soft; and so on. The question is: what determines which of these qualities is the real quality, the one that does belong, as opposed to appearing to belong, to the item itself?

It won't help to say that the real qualities are those which the

## THE ARGUMENT FROM ILLUSION

object has independently of our perception of it, for we would still have to discover what qualities it has when we are not perceiving it. Nor will it do to say that the real quality is the quality that everyone, or most people, perceives when they perceive the object. We have seen (6.1) that an object can appear e.g. red to everyone, because of the presence of a red light, without really being red. This is why I distinguish calling something 'really' red, from calling it 'objectively red'. Nevertheless it seems that what counts as the real quality of the object will have to be explained in terms of what is perceived under certain conditions. The appearance of a thing and the qualities we perceive can vary according to the conditions under which that thing is perceived, and these conditions are of two kinds: those that would affect any perceiver and those that affect only the individual. So we may expect the real qualities of the thing to be defined in terms of what will be perceived given certain environmental conditions and a certain state of the perceiver. This can be done by reference to 'standard conditions' and 'normal perceivers', the real colour of an object, for example, being defined as the colour seen by a normal percipient under standard conditions, i.e. in broad daylight.[1]

This is sometimes described as a 'conventionalist' account, the suggestion being that it is just a matter of convention which of the various colours we see we pick on as the real colour of an object, and this might suggest in turn that it is just a matter of convention, rather than of objective fact, that we talk of things as really coloured at all. But it is hardly a matter of convention that the

[1] It has often been argued (e.g. Price I, p. 20; Ayer VI, pp. 128–9) that this definition is circular because 'standard conditions' and 'normal perceivers' are themselves defined in terms of perceiving the real qualities of things. But this is certainly not the natural way to define these terms. Standard conditions would most naturally be thought of as those under which we do most of our perceiving (or, perhaps, those which obtain where objects of the kind in question are almost invariably to be found. This would mean that Austin's fish (II, p. 65) which is vividly multi-coloured at a depth of a thousand feet but a muddy greyish white in sunlight is really multi-coloured, rather than really greyish white, because that is how it appears when perceived in its natural habitat). And a normal perceiver would most naturally be thought of as one whose perceptuable apparatus is, if not precisely the same as most people's, at least not so significantly different as to cause him to perceive things as different from what most other people perceive them to be. Given this account of standard conditions and normal perceivers, and the above definition of real qualities, it naturally follows that under standard conditions things appear to normal perceivers as they really are. But since standard conditions and normal perceivers are specified without reference to real qualities this does not involve any circularity.

standard conditions for the observation of colours, for example, are that it be daylight, that there be no interfering sources of light, no coloured objects between the perceiver and the thing to which the colour is ascribed, and so on. If we ordinarily saw things under ten fathoms of salt water then no doubt we would think of them as being different colours from what we now think of them as being, but it is not a matter of convention that we do most of our perceiving under daylight, rather than under ten fathoms of salt water.

Even so the fact remains that it is a sort of an accident that we normally see things in daylight, without interfering sources of light, etc., and so it is a sort of an accident that we say that roses are red rather than some other colour which they might appear to have under other conditions of observations.[1] If we all lived in a Platonic cave lit only by sodium lights we would naturally ascribe different colours to objects from those we ascribe now. Then if we came out into the daylight we would be faced with the choice of saying that these red and orange objects appeared many new and vivid colours under these new, special, conditions of observation, or of saying that the objects were really these vivid colours and only appeared red and orange under our old, special, conditions of observation. And, significantly, which we would say would depend on whether the cave turned out to be but one small part of a much wider world which did not have this special (sodium light) illumination, or whether the illumination by daylight turned out to be the exception rather than the rule. In other words which type of illumination we regard as standard, as showing things in their true colours, depends on which type of illumination is the

---

[1] Zeidins (I, § 4) has argued that what constitutes standard conditions for the observation of colours is not, even in this limited sense, a conventional matter. He suggests that the standard conditions are not the most common but those that are the best for distinguishing different colours. But this can hardly be so. Stamp collectors use ultra-violet light to bring out slight differences in the colours of stamps. On Zeidins' account this should mean that the standard conditions for the observation of colours should include ultra-violet light, which means, in turn, that all objects are really shades of violet! The truth is rather that the use of ultra-violet light can lead us to say that two objects which are indistinguishable in daylight are in fact different colours or shades, but it, or any other device, can do this only so long as it does not conflict with the account of what is seen under daylight. After all, if the two objects are green in daylight we will say that the ultra-violet light shows that they are different shades of *green*, even though the colours seen under the ultra-violet light are different shades of violet.

most common. But this 'relativity' of real colours should not tempt us to doubt whether objects are really coloured at all. 'Is it really red?' is quite a different question from 'Is it really coloured?'

This definition of real qualities by reference to standard conditions and normal perceivers can be given for temperature, taste, smell, sound, and texture as well as for colour. That is, it can be given for those qualities which are 'appearance-determined' (cf. 5.1). What colour a colour is is determined by how it appears, and similarly what colour an object is is determined by how it appears under certain conditions. But shape, size, etc., are not appearance-determined qualities, and what shape or size an object really is is not determined by how it appears. Rather it is determined by the use of measuring instruments, such as rulers and protractors. In fact the account we have given, in terms of how things appear under standard conditions to normal perceivers, cannot apply to non-appearance-determined qualities, if only for the reason that with such qualities it is possible for the object to appear to have two different, incompatible, qualities at the same time. The moon looks the size of a sixpence, but also looks the size of a half-crown; the tilted penny looks elliptical, but also looks circular. It is tempting to suggest that all we need do here is specify further the conditions under which the object is to be perceived, e.g. to include the condition that the object must be held at right-angles to the line of vision, in which case it will not appear elliptical at all. One objection is that this addition scarcely counts as a 'standard condition' in the sense of one which applies when most people do most of their perceiving, and so it may seem that this addition is purely arbitrary, making the penny's being circular a conventional matter in the strongest possible sense, i.e. simply a matter of what we prefer to say.[1] But the main objection is that this condition works for only a small range of things, albeit things of the kind philosophers tend to restrict their discussion to. Suppose I am set to discover the real shape of Professor Ayer. Is it really suggested that I hold him at right-angles to the line of vision and so, by noticing what shape he appears under these conditions, ascertain his real shape? Unlike the penny Professor Ayer is not a flat object, so it is not even obvious what counts as holding him at right-angles to the line of vision.

The obvious truth of the matter is that the real shape of an object

[1] Zeidins I, § 2–3 shows this to be a mistake.

—and even more obviously its real size, for under what conditions does an object look its real size?—is established not by how it looks, but by measurement. Of course perception will be involved in trying to discover the results of such measurement, so in a sense in which anything discovered via perception involves discovering how things appear, discovering the real shape of an object involves discovering how it appears. But its real shape is not a matter of how it appears under certain conditions in the way that its real colour is.

As well as the question of what colour, shape, etc., an object is there is also the question of whether it is really coloured, really possesses a shape, etc. It would seem that just as to say that a sound is real is to say that it can exist without being heard, so to say that objects are really coloured, in the sense that interests philosophers, is to say that those objects possess colours whether we perceive them or not, i.e. that their being coloured is a matter quite independent of our perception of them. Nevertheless the fact that the colours we see and ascribe to objects depend, in various ways, upon conditions of observation, may suggest that objects are not really coloured in this sense. There seem to be two main reasons for insisting that they are. First, so long as we accept a Realist theory of perception according to which what we perceive is the object itself it seems unwarrantably strange to try to separate the colour from that which is, or appears to be, coloured. If we adopted a Lockean percept theory of perception according to which what we perceive is not the object but a percept, a sort of picture, of the object then we could say, with reasonable comfort, that the colour is only a feature of the percept, the object as pictured, and not the object itself. But once it is agreed that what is perceived is the object itself there is something very strange about saying that when I see a coloured surface the surface is really there but the colour is not. For in seeing the surface I see the colour, and if the surface had no colour I could not see it. The second reason is that although the colours one sees may vary depending on the conditions of observation, these different colours are connected with the colour that is actually ascribed to the object. Although the colour I see may not itself be the colour belonging to the object, what colour I see is, among other things, determined by what colour the object really is. So since the colour of the object helps determine what colour we see when we look at it, it is natural to say that the object is that colour even when that is not the colour we see; and so to say that that colour really exists, is a feature of the world independent of our perception of it.

## 6.3 APPEARANCE

The topic of appearance, and through it the topic of sense data, can be made very confusing by a failure to distinguish two possible senses of the verbs 'appear', 'look', 'sound', etc.[1] I shall call the first the 'resemblance' sense. To say that something appears, looks, etc., $y$ in this sense is, roughly, to say that, as it is perceived to be, it resembles a $y$ thing, something which is $y$. When we say of the famous tilted penny that it appears, or more idiomatically looks, elliptical we mean, roughly, that it resembles such things as are elliptical. We can also say that the tilted penny looks circular because, of course, it resembles a circular object, one held at an angle to the line of vision. But this would not be the natural thing to say. The natural thing to say is not that it resembles a circular thing but that it *is* a circular thing. It is worth pointing out that 'It appears elliptical', 'It looks red', etc., mean, in this sense, 'It resembles such things as are elliptical', 'It resembles such things as are red'. 'It looks red' does not *mean* 'It resembles tomatoes and phone boxes' for it is not a necessary fact that tomatoes and phone boxes are red. Indeed to say 'It looks red' is not to say that anything is red, for it does not follow from this that there are red things. All that is meant is that this is like what a red thing would be.

I shall call the second sense of 'appears', etc., the 'judgement' sense. To say that something appears, looks, etc., $y$ in this sense is, roughly, to say that the speaker thinks, or is inclined to think, that it is $y$. If I say 'It appears heavy' or 'It looks like rain' I might mean that it resembles a heavy object (or perhaps that the way you are struggling with it resembles the way you would struggle with a heavy object) or that the marks on the window resemble those made by rain. But it is more likely that I mean that I think it will turn out to be heavy or I think it will rain. 'It appears', 'It looks', etc., characteristically have this sense when the 'it' is not a pro-

---

[1] I am here concerned with 'appear' in the sense that is concerned with perception. 'Appear' is not always used in this way, cf. 'It appeared in the newspaper', 'He appeared before the Magistrates' Court'.

noun, does not refer to any particular object. 'It appears that I am mistaken', 'It looks as if they've won', 'It sounds as if they're out' can hardly be met with the question 'What does?'

It is, I think, quite clear that these verbs do have both these senses. 'Whisky looks like water', 'The soprano saxophone sounds like an oboe', 'The penny looks elliptical' (when I know quite well that it is not), 'He looks like my father' (when I can see quite well that he is not) use 'looks' and 'sounds' in their resemblance, and not their judgement, sense as clearly as 'It looks as if they've won' or 'It sounds as if they're out' use them in their judgement, and not their resemblance, sense.

It is important to keep these senses distinct but it may be equally important to realize that these verbs are not always used unambiguously in one sense rather than the other. Many statements like 'It looks heavy' can be interpreted in either of the two ways, and if it makes no difference which way we interpret it, there is no point in insisting that it be one rather than the other. Nevertheless it will often be worth our while to draw a strict, if sometimes arbitrary, line between the two senses, and so avoid the shift from one sense to another. For, as we will see, it is only when the judgement sense is involved that statements about the look or appearance of something are incorrigible.

What both senses have in common is that statements about how a thing appears are essentially non-committal. The point of saying 'It appears heavy' or 'It looks red' is to avoid committing oneself to a judgement as to what it really is. This has led some to think that we should only use such judgements when we are in doubt as to what the thing really is, or where we know that it is not as it appears to be.[1] I think the point is that if one knows that $X$ is $y$ then there is something odd and misleading about restricting oneself to the weaker claim that $X$ looks or appears $y$, just as there is something odd and misleading about a statement like 'Some members of the Houses of Parliament speak English'. But this oddness and misleadingness doesn't mean that the statement is mistaken or illegitimate; it is both meaningful and true. '$X$ appears $y$' naturally *suggests* that the speaker isn't sure, or that $X$ is, in fact, not $y$, just because 'appears' is non-committal, but it doesn't *mean* this. After all we can say such things as 'It's expensive and it looks it' or 'He's not just a musician, he even looks like

[1] Grice I argues in detail against this.

one'. The two senses are also related in that if a thing appears (resemblance sense) *y* then a perceiver who does not know any better, who does not have any other information to go on, will naturally be inclined to judge that it is *y*, to say that it appears (judgement sense) *y*.

'Seems', on the other hand, is used only in the judgement sense. It may be because a thing looks (resemblance sense) elliptical that I say that it seems elliptical, but what I mean by the latter is not that it resembles elliptical things but that I am inclined to judge that it is elliptical. If I know or can see clearly that it is not elliptical then I should not say that it seems elliptical, but only that it looks elliptical. We do not say 'Whisky seems like water' or 'He seems like my father' (unless we mean that we think the whisky is like water, that he is like my father) as we do say 'Whisky looks like water' or 'He looks like my father', and we do not talk about 'seemings' as we do talk about 'appearances'.

The noun 'appearance', then, naturally goes with 'appears' in the resemblance sense. There might be cases where 'appearance' has a sense connected with the judgement sense ('keeping up appearances', perhaps, or 'He gave the appearance of being satisfied') but on the whole it naturally goes with talk about the look, sound, taste, smell or feel of things, where we have in mind not so much what we think those things are like as what they actually are like, i.e. what they resemble. Notice that this is not to say that a thing's appearance is what it looks, sounds, etc., *like*. The white object that looks blue through blue glass is blue in appearance, not white in appearance, even though it looks like a white object (as seen through blue glass). If I say of a cloud that it looks like a whale I do not mean that it is a whale in appearance or that it has a whale appearance, I mean that it is *like* a whale in appearance, that it has a whale-like appearance.

Finally the sense of 'appears' which we used in explaining what it is for a quality to be 'appearance-determined' was, of course, the resemblance sense.

## 6.4 ILLUSION

Appearance, hallucination, illusion, delusion, non-veridical perception—philosophers have often tended to use these terms interchangeably. It is perhaps the chief merit of Austin's *Sense and Sensibilia* to point out that these things are not equivalent and that it is only if we do not distinguish them that certain arguments—the argument from 'illusion'—will seem plausible. We have, then, to distinguish these things.

One feature of sensory illusion[1] is that the thing perceived appears other than it really is, but this cannot be the whole story for the tilted penny appears other than it really is without us wanting to call it an illusion. A second condition for illusion is that as well as looking what it is not, the object must also not look what it is. In Austin's example the 'headless woman' constitutes an illusion because not only does it look as though she has no head, but also it does not look as though it is a woman with her head in a black bag standing in front of a black screen. Nevertheless, a white object seen through blue glass looks blue and does not look white, yet we would not want to call that an illusion either. Perhaps we need the third condition that we actually be misled by appearances. But that will not do for the 'headless woman' and the Muller-Lyer illusions remain illusions even when we know full well that this isn't a headless woman, that one line is not longer than the other. And even if I am misled into thinking that the object is blue I don't think we would call seeing the white object through blue glass an illusion. Perhaps the third condition should be something like that there be no way of telling, from present perception, that the object is not as it appears to be. After all when I look at the white object through blue glass everything looks blue and this is one way of telling, or at least coming to think, that the object is not blue as it appears to be, and even what colour the object really is. This would also explain a certain hesitancy about the example of the stick in water which looks bent. Is this an illusion or not? If the suggested third condition is correct then the answer depends on whether the presence of water, which makes other differences to how the stick looks, is thought of as providing a way of telling that the stick is not bent as it appears to be. Perhaps whether or not we talk of illusion depends on how familiar the individual is with the particular phenomenon.

Illusion, then, is different from a thing's appearing different from what it really is—the penny's appearing elliptical does not constitute an illusion. Nor does all illusion involve non-veridical

[1] I am not here concerned with 'illusion' in the sense of being mistaken where this has nothing to do with perception (cf. 'being under an illusion') although these senses are closely connected.

perception; the Muller-Lyer illusion involves non-veridical perception only if I am actually misled by appearances into thinking that one line is longer than the other. Nor, for that matter, are all cases of a thing's appearing other than it is cases of non-veridical perception; my perception of the white object through blue glass is only non-veridical if, once again, I am actually misled by appearances into thinking that it is elliptical. And, of course, illusion is quite different from hallucination. Macbeth did not suffer an illusion nor does the man who sees the headless woman suffer a hallucination.

Delusion, too, is different from all these things. The penny's appearing elliptical, or the illusion of the headless woman, do not constitute delusion. Unlike illusion delusion necessarily involves mistake, and, as Austin says,[1] it is restricted not just to cases where we are mistaken but to cases where something has gone radically wrong. I would myself be inclined to speak of delusion only in those cases where a man mistakenly thinks he is perceiving something, not just in the mild sense of perceiving something which he thinks is something else, but in the much more extreme sense of thinking he perceives something when he perceives no such thing at all.

Why have philosophers always tended to run these different things together? The main source of confusion is, I think, the fact that 'appears' has those two different senses. The fact that something appears (resemblance sense) to be other than it is is taken to mean that it appears (judgement sense) other than it is, which means that the perceiver takes it to be what it is not. Thus the fact that something appears other than it is is thought of as involving delusion and illusion (in the sense of making a mistake), and as making the perception non-veridical. And once we identify non-veridical perception with having hallucinations the confusion is complete.

One final comment: this discussion of reality, appearance and illusion has, throughout, been carried on with Realist assumptions. If the Idealist, in particular, wants to make use of these notions he may have to give them a rather different sense from that I have claimed for them. My justification for restricting myself to the Realist interpretation is that ordinary language, from which these terms are borrowed and in which their philosophical uses are

[1] II, p. 23.

grounded, is inescapably Realist. If Realism is rejected then these ordinary notions will have to be altered.

## 6.5 THREE ARGUMENTS

We come, at last, to the argument from 'illusion', the argument most commonly invoked in the attempt to show that what we perceive are not, as the Realist claims, external objects but sense-dependent percepts. In view of the confusions between hallucination and illusion and appearance and delusion and non-veridical perception it is not surprising that this heading 'The Argument from Illusion' should cover more than one argument. I want to discuss three lines of argument which I will call 'The Argument from Hallucination', 'The Argument from Appearance' and 'The Argument from Qualitative Similarity'.

The argument from hallucination is quite simple. It appeals to the fact that people sometimes perceive things which do not really exist and concludes that 'Naive' Realism must, therefore, be false. 'Though some sense data may be parts of the surfaces of objects, some certainly are not. So Naive Realism is, after all, still false'.[1] If Naive Realism is supposed to maintain that *all* perception is perception of external objects then not only is it obviously false, but also it is much more naive than any common sense theory of perception. Common sense Realism ('Direct' Realism, if you prefer) is the theory that not all perception is perception of percepts, that most perception (including all veridical perception) is perception of external objects.

The argument from hallucination may be linked to the argument from appearance via the troublesome case of double-vision. What does a man with double-vision see two of? Obviously not the external object, for there is only one of them, so it must be something else that he sees, a percept existing only in so far as he perceives it. But this question is 'a trick question, committing the Fallacy of Many Questions. It assumes that he sees two of something when he does not—he sees one thing looking double'.[2] The simple fact is that, under certain conditions, things can and do look double—or, if you like, we can and do see the same thing twice. And even if we want to say that one, or both, of the images exists only in our perception of it this does nothing to prove that everything we perceive is sense-dependent.

[1] Price I, p. 63.   [2] Hirst I, p. 49.

This leads us to the much more common phenomenon that things can, and in a way usually do, look different from what they really are. It is on this fact that the argument from appearance relies. We have seen that a white object can look blue and a circular penny elliptical, although two points need to be remembered: first that if a thing looks or appears different from what it really is this does not necessarily mean, except where appearance-determined qualities are concerned, that it cannot at the same time look or appear as it really is (the penny may look both elliptical and circular); and second that a thing looks or appears such and such does not necessarily mean that the perceiver notices that it looks or appears such and such—most of the coins I see look, in the appropriate sense, elliptical, but it is seldom that I notice this.

The argument from appearance can be stated:
(1) This penny is circular.
(2) What I perceive is elliptical.
(3) Therefore what I perceive cannot be the penny and must be something else.

The argument is also, perhaps more frequently, stated in the form that since what different people perceive is different (the penny is elliptical for one, circular for another) they cannot be perceiving the same thing. In each case the argument is valid but the second premiss is obviously false—or else begs the question. For the Realist it is simply incorrect to say that what I perceive is elliptical. The truth is that it looks or appears elliptical, and no contradiction is involved in saying that something looks elliptical but is circular. Indeed if it didn't look elliptical in this way it could hardly be circular.

It is quite astonishing how often reputable philosophers from Berkeley to Ayer have confused 'looks' with 'is' in this way. Two factors explain this simple and obvious mistake. The first is the tendency to reify appearances, to take it for granted that what is perceived is not the thing, which looks elliptical, but the appearance, which *is* elliptical. Of course to assume this, that we see not the circular penny but an elliptical something else, is simply to beg the question. The second factor is the use of the expression 'what is perceived' which may refer either to the object that is perceived or the quality that is perceived. When I see the white object through blue glass the object I perceive is white but the colour I perceive is blue. It is easy to slip from saying that what

I perceive (the colour) is blue to saying that what I perceive (the object) is blue. But the object is not blue; it merely looks blue.

It is worth adding that the fact that the colour I see is not the colour of the object does not mean that I cannot be seeing the object. Using the distinction between identification and description (6.2) we can say that although 'A blue wall' is a perfectly accurate description of what is perceived it would be incorrect as an identification. We can even say 'This red wall is a white wall' without contradiction, so long as the former phrase is a description and the latter an identification of what we perceive. Or we might point out that in the sense in which 'I perceive a red wall' is true it is quite compatible with 'I perceive a white wall', in the sense in which it is true. This very example brings out that there is nothing odd in the fact of a thing's being perceived as different from what it really is, and that there is no contradiction between 'The external object which is perceived by me is white' and 'I perceive the external object as being red'.

Our rather cavalier habit of joining identifications and descriptions together can be a major source of confusion. 'The sheep was nothing but a white dot on the hillside' or 'The boat became a black speck on the horizon' can seem absurd. Of course the sheep was more than a white dot; obviously the boat was not transformed into a black speck. What we mean is that from the particular point of view the sheep appeared to be, looked like, a white dot, that the boat came to look like a black speck, and there is nothing puzzling about these remarks. Again when we say 'That dot on the hillside is a sheep' or 'That speck on the horizon is a boat' we are identifying the dot or speck as (*not* with) a sheep or a boat. 'That dot' and 'That speck' are descriptions, not identifications, of what we perceive, they are descriptions of what we perceive from this point of view.

Finally we have the argument which contains the basic truth in the traditional argument from illusion, the argument from qualitative similarity. This argument often depends in a confusing way on the argument from appearance, as when it is argued that as I walk around the table what I see changes continuously in shape and only in the one case is the shape I perceive the real shape of the table. In all other cases what I perceive cannot be the table; it must be something else, a percept. Since there is no dramatic change in what I perceive as I move around the table it seems implausible to suggest that what I perceive on one occasion really exists, while what I perceive on all other occasions is a sense-dependent percept. In this common formulation the argument is easily avoided by pointing out that it is false to say that I perceive different things from these different points of view. The truth is that I perceive one thing which happens to look different, depending on my position.

If this argument is to be given a decent run for its money it must be stated not with reference to the fact that things can and do look different from moment to moment and place to place, but by reference to hallucinations and the like:

(1) All the things we perceive via a particular sense-modality are qualitatively alike.

(2) So it seems natural and preferable to say that all the things we perceive are of the same ontological type.

(3) But some of the things we perceive, e.g. hallucinations or after-images, are sense-dependent percepts.

(4) So it seems natural and preferable to say that all the things we perceive are sense-dependent percepts.

One way of attacking this argument might be to point out that when what is perceived does not really exist it usually is noticeably different from what we perceive when it does really exist. Hallucinations and after-images are not exactly like ordinary tables and chairs. However all that the argument needs is the logical possibility that things with perceived existence can be qualitatively indistinguishable from that which really exists. That is, there is no *essential* qualitative difference, so far as our perception on any particular occasion goes, between the external objects and the percepts we perceive. This raises the epistemological problem of how we know whether what we perceive really exists, but it does nothing to show that we cannot know this, let alone prove that what we perceive does not really exist. Perhaps one source of error here is the common assumption that perceiving is a form of knowing (cf. 11.5). Given this it is easy to say that since we can perceive something without knowing whether it really exists, it follows that what is perceived does not really exist.

However the main point of the argument from qualitative similarity is that it presents us with an alternative. We can either allow that things that are, or at any rate could be, qualitatively alike are nevertheless ontologically distinct, or agree that we never perceive external objects, that we always perceive percepts. There seems no special reason for adopting the second alternative, which is made attractive mainly by exaggerating the number of occasions when we would naturally say that what we perceive exists only in so far as it is perceived, i.e. by suggesting that whenever what we perceive looks different from what it really is we must be perceiving something which exists only in so far as it is perceived. The argument from qualitative similarity is not conclusive, but it may be an influencing factor when we come to choose between the theories. In so far as we want to agree that things that are qualitatively alike will also be ontologically alike so far we will be inclined—although not forced—towards a percept theory of perception.

CHAPTER 7
# THE ARGUMENTS FROM SCIENCE

### 7.1 THE CAUSAL ARGUMENT

The Causal theory has always derived most of its support from the facts of physics and physiology. These facts give rise to two rather different arguments which I will distinguish as 'The Causal Argument' and 'The Argument from Physics'. I begin with the former.

Investigation has shown that when we perceive things complicated and often quite lengthy chains of causally connected events are involved. Seeing an external object involves light-waves of various frequencies being reflected by the object and impinging on the retina, changes occurring in the rods and cones of the retina, an electrical impulse being passed along the optic nerve to the optic centres of the brain, and the consequent stimulation of those centres. It is tempting to think of the seeing as a further causal consequence of these goings-on, as, perhaps, something produced in the mind by the stimulation of the optic centres of the brain. And this, in turn, is taken to show that what we perceive is not, as we might think, the object itself, but rather an effect of various processes involving the object. It is not always held that this effect is a picture resembling or reproducing the external object, indeed the point of the argument from physics is that it does not. Rather this claim distinguishes Representative from non-Representative versions of the Causal theory.

Several objections have been raised against this argument, this interpretation of the physical and physiological facts, but although some count against the theory I doubt whether any show it to be mistaken. The standard objection is that the argument is, in effect, self-refuting, in that it makes the very facts on which the argument and the theory are based unknowable. For the conclusion, that we always perceive percepts and never perceive physical objects, means that we can have no knowledge of those various causal processes to which the argument refers. In fact we cannot even know of the existence of external objects in the first place! 'In short, if there were external bodies, it is impossible that we should ever come to know it; and if there were not, we might have the very same reasons to think that there were that we have now'.[1] It seems that if we are to adopt a causal theory we must insist, with Kant, that the causes of our percepts are completely unknowable. But then how can we appeal to facts about these causes in order to establish that what we perceive is not the object but some effect of it?

The Causal theorist can avoid this difficulty only by treating external objects as theoretical entities whose existence is to be inferred from, and as an explanation of, what we perceive. Our ordinary talk about external objects has to be regarded as an explanatory hypothesis invoked to explain how and why we perceive the percepts we do. This makes the causal argument more difficult to state—we have to begin with facts about our percepts rather than with scientific facts about external objects and our sense organs—but I do not see that it makes the argument, or the theory, incoherent. The argument is oddly circular, in that we establish the scientific facts by reference to our percepts, and then establish that we perceive percepts by reference to the scientific facts, but I don't see that the circle is vicious.

However it has been argued that the causal argument is not just circular but self-contradictory, in that it begins by citing facts that are involved in perceiving external objects and ends by denying that we perceive external objects at all.[2] I think the objection can be avoided. The Causal argument does not need to assume that we perceive external objects. We might distinguish 'direct' from 'indirect' perception, saying that to perceive something 'indirectly' is to perceive not that thing but some effect of it. Thus to hear a waterfall is to perceive the waterfall not directly but indirectly. I do not hear the water itself but only the sound it makes, i.e. I hear the water only in so far as I perceive its effect. The Causal theory is the theory that we never perceive external objects directly but only indirectly, i.e. we perceive them only in the sense that we perceive their effects, the percepts they produce in our minds. So the Causal theorist can say that we see cars just as we hear waterfalls, i.e. in the sense that we perceive their effects. It is in his interpretation of our ordinary perceptual statements, rather than in a rejection of them, that the Causal theorist departs

---

[1] Berkeley I, § 20.  [2] Cf. Hirst I, p. 172.

from common sense. This move also enables the Causal theorist to avoid the charge[1] that his account is nonsensical in that it denies that an object like an orange is visible, tangible, etc. The Causal theorist allows that such objects are perceivable, although they are not perceivable in the sense or way we might normally think they are.

Another argument, directed in particular against Representative theories, is the Berkelean one[2] that the theory makes it impossible for percepts to resemble external objects, as the theory itself claims they do, since percepts are perceptible and external objects, according to the theory are not. Is this more than a re-iteration of the point that the theory makes it impossible to tell whether our percepts do, as it claims, resemble external objects? Some have thought so: 'The characteristic $X$ which is a characteristic of a certain sense-impression . . . must be an immediately perceivable characteristic, because sense-impressions are immediately perceivable. Now, by hypothesis, no characteristic of physical objects are immediately perceivable, therefore no physical object can have the characteristic $X$'.[3] This argument rests on a confusion between qualities, as universals, and particular instances or examples of those qualities. We might as well argue that since the coal in the blast-furnace cannot be perceived it cannot possess perceptible qualities, such as colour. Perhaps a particular expanse of, say, red, must be immediately perceivable because it belongs to a 'sense-impression', but that doesn't mean that all expanses of red, all instances of red, must be immediately perceivable.

There is also the argument that since 'immediate and mediate perception are correlative terms . . . we can understand talking of the one only if it makes sense to talk about the other. Now if physical objects are mediately perceived, as the Representative theory asserts, then we can only understand this assertion if it makes sense to talk of their being immediately perceived'.[4] We might as well argue that since 'divisible' and 'indivisible' are correlative terms we can understand talking of the one only if it makes sense to talk about the other, and hence, since a geometrical point cannot be said to be divisible it cannot be said to be indivisible either. But, of course, we can call a geometrical point indivisible, and the point is that it cannot, logically cannot, be divided.

[1] Cf. Warnock I, p. 178.
[2] Cf. I, § 8.
[3] Armstrong I, p. 31.
[4] Armstrong I, p. 34.

This is legitimate so long as both sides of the divisible-indivisible distinction have some application somewhere, but not necessarily both in this particular case.

For my own part the main fault in the causal argument, and consequently in the Causal Theory itself, is that it involves a serious misinterpretation of the scientific facts on which it relies. To say that *perception* involves causal processes, and even that it is caused by or is causally dependent upon the existence and presence of the external object we say we perceive, is not to say that *what is perceived* is caused by or causally dependent upon the external object. This is, in fact, the familiar mistake of confusing perception with what is perceived, a confusion which has been particularly obvious in the writings of Bertrand Russell. For example, Russell confesses himself 'surprised to find the causal theory of perception treated as something that could be questioned. . . . A gun is fired, let us say, and people are ranged at various points 100 metres, 200 metres, 300 metres and so on, distant from it. They hear the noise successively. This evidence would be considered amply sufficient, but for philosophic prejudice, for the establishment of a causal law making the hearing of the noise an effect of a disturbance travelling outward from the gun'.[1] But, obviously, the undeniable fact that hearing the noise is the effect of this disturbance does nothing to show that what is heard is such an effect. 'The argument . . . confuses the *mechanism* on which the perception is dependent with the *objects* that are perceived'.[2] It also gives rise to many oddities; perceiving these objects requires neither eyes nor brain—these are causally responsible for what we perceive but are not used to perceive, according to the theory!

Science has shown that various causal processes are responsible for, are the causal conditions of, our perception. The mistake has been to think that these causal processes involved in perception are responsible for, are the causal conditions of, what we perceive. The theory is that stimulation of the retina, etc., produces a visual percept in our minds, but quite obviously what these processes produce is not what we see but the perception itself. The end result of these causal processes is perception—what we perceive is the thing, whatever it is, that comes at the beginning, the thing that reflects the lightwaves onto our retina. This point holds good whether the theory is that we perceive something produced 'in

[1] VII, p. 702.  [2] Mundle I, pp. 70-1.

## THE ARGUMENTS FROM SCIENCE 117

our mind' by the causal processes, or whether the theory is that we perceive something produced in our brains. Russell[1] and Hirst[2] have identified the activity in the optic centres of the brain not with the cause of what we see, but with what we see itself. But clearly if mental states are to be identified with brain states, the obvious thing to identify the activity in the optic centres with is the mental process of perceiving. If we were to identify this brain activity with that is perceived. we would then be unable to find brain activity to identify with the process of perceiving! There are good reasons for identifying the brain activity with the perceiving (e.g. that perceiving is impossible without that brain activity) but none at all, that I can see, for identifying it with what is perceived. The fact that the brain state may vary as what is perceived varies provides no reason for saying that the brain state is what is perceived, for if we think of the brain state as the registering of what we perceive the brain state will be expected to vary as what is registered varies. But this doesn't mean that what is registered is identical with the registering, for then there would be nothing to register.

In short, so far from resting on the physical and physiological facts the causal theory seems to me to rest on a gross, and im-

---

[1] Cf. V, p. 146.

[2] I, although I am rather hesitant about ascribing this theory to Hirst, as I find it difficult to understand what precisely his theory is. It seems to be an incoherent mixture of Realism and the Causal theory, according to which what we perceive is objective, independent of the perceiver, and yet at the same time merely an 'adverbial' aspect of the perceiver's perception, something produced in us by the action of external objects on our sense organs. Hirst argues in detail against the traditional Causal theory, and insists that he differs from those who say that the brain activity causes a percept which we then perceive, but fails to see that his own version of the theory, in which what we perceive is identified with the brain activity, is subject to precisely the same objections (e.g. those on pp. 172-3 of his book). Nor can I see why he finds Realism unsatisfactory. He says (pp. 319-20) 'As soon as one tries to understand how perception occurs... one is forced by the scientific evidence of causal processes and the physiological evidence of modificatory processes, into the Representative theory or some theory like mine on which we perceive an object by having an experience caused by its acting on our sense organs'. But common sense Realism is just such a theory, a theory according to which we perceive an object by having some experience caused by its acting on our sense organs. Of course a failure to distinguish what is perceived from the perception of it may lead us to misinterpret that troublesome word 'experience', and so think that what is required is a theory according to which perceiving an object is perceiving an 'experience', a percept, caused by the object's acting on our sense organs. But that would be a mistake.

plausible, misinterpretation of those facts, a misinterpretation due, in the end, to a failure to distinguish between perception and what is perceived. It is the former which is causally produced in the brain or mind; not the latter. There remains one point in favour of the causal theory that has still to be considered; this is the Time Lag argument. Light from a star may take many years to reach the earth. In the meantime the star may have exploded, so that when we look into the night sky the star that we would say we saw no longer exists. Surely this shows that we cannot be seeing the star, but only some effect of it? As Russell puts it[1] to see the light from the star is no more to see the star than to see a New Zealander in London is to see New Zealand. It is obvious that such examples show that our naive common sense beliefs about what we perceive will have to be altered in some way. We might, without too much trouble, say that in fact we do not see stars at all, except in so far as we see the light from them, and saying this would not tempt us to say that we never see tables or chairs either, but only the light from them. There is a sense in which we do not see the star but only its light, but do see the table and not its light. Or we could even say that, as this example shows, we can see 'into the past', i.e. see what no longer exists. This is certainly contrary to naive common sense, but once we understand the facts about the finite velocity of light we can say this quite happily without forcing ourselves to the conclusion that we never see things themselves but only their effects.

## 7.2 THE ARGUMENT FROM PHYSICS

We now turn to the second argument from science, the argument from physics. This relies on the alleged fact that science has shown that the external world is not at all as we perceive it to be, and therefore that what we perceive cannot be parts of the external world. It is worth noting that this argument is incompatible with any Representative theory, as its whole point is to show that what really exists is not at all like what we perceive. The argument is that I perceive a solid, coloured, stationary table but what is really there is a discontinuous mass of rapidly moving non-coloured items, so what I perceive cannot be what is really there. Similarly I perceive a sound but what is really there is a certain motion in the air. As Russell has put it:[2] 'Naive Realism leads to physics,

[1] V, p. 144.    [2] VI, p. 15.

and physics, if true, shows that Naive Realism is false. Therefore Naive Realism, if true, is false; therefore it is false'. This argument can be turned, with equal force, against physics: if Naive Realism is false then physics is derived from false assumptions. Admittedly this does not show that physics is false, but it does seem to destroy all reason for regarding it as true. This in itself is sufficient to suggest that something is wrong with the argument from physics.

What is wrong is the suggestion that when the physicist tells us that this is a collection of atoms, neutrons, molecules or whatever, he is denying that this is a table. For the collection of atoms, or whatever, just is a table. I do not mean that 'table' *means* 'collection of atoms', but it is a matter of scientific fact that the table has this constitution, is, among other things, a collection of atoms. Similarly it is a matter of scientific fact that the sounds we hear, or at any rate the real non-hallucinatory sounds, are constituted by sound-waves, motions in the air. This is why the suggestion that physics is inconsistent with Naive Realism counts as much against physics as it does against Realism. The physicist's aim is to discover what, at a microscopic or sub-microscopic, and therefore sub-perceptual, level, the table, the macroscopic perceptual object, consists of, to discover what, physically speaking, it is made up of. To turn round and deny that there is a table there would be to make nonsense of the whole programme.

Even so, it may be said, the physicist can avoid this absurdity by insisting that there is a table there, but adding that the table is nothing like what we think it is: it is discontinuous not solid, full of moving parts not stable, colourless not coloured, and so on. To say this is, in part, to confuse the table with the atoms of which it is said to be constructed. What I mean by calling the table solid is that if I try to walk through the place where it is I will feel something preventing my motion, and that no matter how closely I examine it with eyes or fingers I will not notice any breaks or gaps in it; what I mean by calling it stable is that no matter how closely I examine it with eyes or fingers I will not discover any moving parts; and so on. To point out that the table consists of imperceptible elements which do move around and which are separated by gaps is not to deny any of this. If by calling the table solid we mean that no matter how closely and by whatever means we examine it we will never discover any parts separated by gaps, then physics has shown that the table is not solid. But it is rather exaggerated to suggest that this is what is meant by calling the table solid; all that we mean is that so far as ordinary perception and observation goes no gaps will be found, and physics has not shown that to be false. Or perhaps we might rather say that originally, when we called things solid, we did not distinguish the claim that there are no perceptible gaps from the claim that there are no gaps at any level. The discoveries of physics have shown that we have to distinguish these two. And it is only when we mean the former rather than the latter that it is true that the table is solid. The investigation of the physicists have not changed the nature or construction of the world, so we accept the modification, restriction, or more accurately clarification of our original ways of talking. Physics has not shown that this way of talking is false, but it has shown more clearly the precise way in which it is true.

### 7.3 THE NATURE OF EXTERNAL REALITY

These considerations raise an important point which lies behind much of the discussion of the last four chapters, and which needs to be brought out into the open if we are to understand fully the nature of a Realist theory of perception. This point can be put by saying that our conception of the external world is fundamentally and essentially *sense-relative*. That is, it is because we have the senses we do, because we perceive the world in the way that we do, that we conceive and know of the external world in the way we do. At first sight this is trite and obvious, but a closer examination shows that this provides the source of both Phenomenalist and Causal theories, and, in part, the distinction between primary and secondary qualities, and the eventual Idealist assertion that all these things exist only in being perceived.

Let us agree that physics has shown that ultimately, in the last analysis, the world is not as we perceive it to be. It is important to remember that this does not mean that the world is not really as we perceive it to be. What it does mean is that when we seek to discover the basic components of which the world is made up, we discover items, like atoms, neutrons, molecules and the like, which we do not perceive and which are in some ways very unlike the things we do perceive. To say that, ultimately, this liquid consists of hydrogen and oxygen is not to say that it is not really water; 'non-ultimate' does not mean 'unreal'. Next, what we perceive, the world as we perceive it to be, is, in effect, a result of relationships, interactions and so on between the various entities of which, ultimately, the world is made up and our various sense organs. It is not at all difficult to imagine a being equipped with different senses, or whose senses are affected by different phenomena from those that affect ours, and what such a being would perceive, the world as he would perceive it to be, would be very different from what we perceive. All of this seems to me indisputable; it is only because we have the senses we do that we perceive such things as tables and chairs, sounds and smells, in the way that we do.

We must not let this mislead us. We may be told that since that there is a table here is just a matter of what we humans perceive, Phenomenalism is, in spite of all, true; or that the fact that what we perceive depends on what senses we have supports the argument from illusion in one or other of its forms; or that the fact that colours and sounds are just the way various phenomena appear to our senses shows that they are are sense-dependent; or that the fact that what is perceived depends on our senses shows that the atoms, light-waves and whatnot cause what we perceive. We are presented with a picture something like this: the world consists of various physical phenomena, atoms, light-waves, etc., which, as such, we do not perceive. What happens is that these phenomena affect our sense organs in various ways and as a result we perceive tables and chairs, shapes and colours, sounds and smells, etc. The suggestion is that these physical phenomena are what is really there, while the tables and chairs, sounds and smells, are but mental entities, a result of the constitution of our perceptual apparatus in relation to the real constitution of the external world. I suppose if someone wants to interpret the facts in this way we cannot prove him wrong, but we must not be bamboozled into thinking this the only interpretation. Particularly in view of our discussion of the argument from physics it seems more sensible to reject this radical distinction between the physical phenomena and the objects we perceive, and to speak instead of the physical phenomena as somehow constituting or making up the objects we perceive. This is, in effect, to identify what we perceive with the physical phenomena, to think of what we perceive as those phenomena as they appear to the relevent senses, as they are perceived to be. There need be no numerical distinction between the sounds we hear and the sound waves the physicist investigates; the sounds we hear are, simply, those sound waves as they are perceived via the sense of hearing, as they are heard to be. Similarly there need be no rigid distinction between the table we see and feel and the atoms the physicist investigates. The table we see and feel is those atoms as they are perceived via the senses of sight and touch, as they are seen and felt to be. Our senses are not sharp enough to perceive individual atoms, but they are sharp enough to perceive various collections of atoms, and to perceive a table is to perceive one such collection.

A being with different sense organs would perceive these same phenomena in different ways, i.e. he would perceive a very different, qualitatively, world. Yet ultimately he would be perceiving the same things as us; in the sense in which we might be said to perceive atoms, light-waves, etc., he would perceive the same atoms, light-waves, etc. But since what he perceives appears so different, the question arises whether we are to say that he perceives the same things in an even stronger sense, whether he perceives the tables and chairs, shapes and colours, sounds and smells that we do (*ex hypothesi* he does not perceive them *as* we do). With sounds for example, I do not think we would say that he does. We feel inclined to insist that sounds as such can be perceived only via the sense of hearing, *our* sense of hearing, and that although the physical phenomena, the sound waves, that are heard as sounds can be investigated and even perceived via other senses (we feel the vibrations of the radio; we see the waves plotted on an oscillograph) what is investigated and perceived in these cases is not so much sound as sound waves. That is, by 'sound' we mean not merely 'sound waves' but rather 'sound waves as perceived via the sense of hearing'. A being that, for example, had its visual apparatus stimulated by sound waves, would not be said to perceive sounds although it could, in the appropriate sense, be said to perceive sound waves. This holds good, I think, for all the secondary qualities; it is a consequence of the fact that they are 'appearance-determined' qualities.

But what of the primary qualities, and physical objects which possess both primary and secondary qualities? Imagine a being who is, somehow, auditorally sensitive to light-waves so that he 'hears' the table where we see it. Do we say that he perceives the table, the physical object that we perceive? We may say that he does perceive the table but because of his strange sense organs it does not appear to him as it appears to us; or we may say that although, in the appropriate sense, he perceives the same collection of atoms as we do, he does not perceive the same table because he does not perceive a table at all. The question is whether or not a word like 'table' is essentially linked to certain human senses, in particular vision and touch, in the way that 'sound' is essentially linked to hearing. I doubt whether there is a clear answer to this question. Our coming across this strange being would force us to refine our linguistic habits, but it is not obvious in advance how we would refine them. I myself am inclined to the view that we would not say that he perceived the table, on the grounds that we can know what tables are without knowing anything about their ultimate physical construction, so that what we mean by 'a table' is the sort of thing I now perceive as I now perceive it to be, and not what is there and may be perceived to be of a very different sort.

Even if we do decide to say that the table is the physical phenomena as they appear to us, with our senses, and therefore that this strange being does not perceive the table, although what he perceives consists, ultimately, of the same elements, this is not tantamount to distinguishing what we perceive from what is really there. Or rather a distinction is made, but it is not a distinction between numerically different and separate items. What we perceive, the table, is still an external object. Our conception of it, what we think of that external object as being, is, unquestionably, sense-relative: it is because we have the senses we do and perceive as we do that we think of the external object as being this sort of thing, a table with these qualities. And since this is the sort of thing we perceive when we perceive the external object, we are correct in describing that object as a table with these properties, meaning that this is the sort of thing we perceive the external object to be. An external object as it appears to us, with the senses we have, a physical object as we perceive it to be, is still a physical, external object.

This brings us to the common suggestion that sensible qualities are 'dispositional' properties, properties which consist in a tendency, liability or power to produce certain effects or results given certain conditions. A paradigm example of a dispositional property is the property of being dazzling which is ascribed, for example, to the beam of a searchlight. The beam is dazzling, even though no one is actually being dazzled by it, and what we mean by calling it dazzling is that anyone who was caught in it would be dazzled. Similarly, it is suggested, the property of being red is a dispositional property, in that to say that a thing is red is to say that anyone who looks at it, under the appropriate conditions, will see the colour red, even if no one is at this moment looking at it under those conditions. But it is not so often realized that all this can equally well be said about physical objects. To say that there is a rug in the next room is as much to say that if someone looked at it they would see a rug, as to say that it is red is to say that that is the colour that will be seen if we look at the rug. What we think of as the external world, the world of tables and chairs, shapes and colours, sounds and smells, is the world *as we perceive it to be*, and not the world as some other being with quite different senses might perceive it to be, nor even the world as it is independent of these different ways of perceiving it.

All this may look like Phenomenalism, even a pure Idealism; it is neither. The Idealist maintains that the things we perceive do not exist independently of our perception of them. I am saying that the things we perceive do exist independently of our perception of them. But in thinking of those things as tables and chairs, shapes and colours, sounds and smells, we are thinking of them as they are, or would be, perceived to be. It is also possible to think of them in other ways—as collections of atoms, for example—even to some extent, to think of them as they might be perceived by some other being. There is an enormous difference between saying that when we talk of a table existing unperceived we are thinking of that external object as we would perceive it to be, and saying that the table exists only in so far as we perceive it. This is the difference between the present position and Idealism. Again, the Phenomenalist maintains that to say that the table exists unperceived is just to talk about what can or could be perceived under certain conditions. I am saying that in calling what exists unperceived 'a table' we are thinking of it as we would perceive it to be. There is an enormous difference between saying that when we say a table exists unperceived we mean that what is unperceived is something which if perceived would be perceived to be such and such, and saying that when we say a table exists unperceived we mean that it can or could be perceived. This is the difference between the present position and Phenomenalism. Even so it is easy to see how this fundamental fact that our conception of the external world is sense-relative could be misinterpreted, and so mislead us into adopting Phenomenalism or Idealism.

# CHAPTER 8
# THE DEFENCE OF REALISM

## 8.1 THE EPISTEMOLOGICAL ARGUMENTS

Our choice between the theories of perception will, to some extent, be guided by epistemological considerations, that is by the extent to which they raise or solve various epistemological problems. In fact the problem of how we can know about what we do not perceive is one of the mainsprings of the Idealist theory that what is perceived exists only in so far as we perceive it. But, in the last analysis, I doubt whether these considerations are wholly on the side of the Idealist.

We have already (7.1) encountered the epistemological argument against the Causal theory, that it seems to make it impossible that we should ever know anything about the existence or nature of the external objects which are supposed to lie behind our percepts. The Causal Theorist has to treat talk about external objects as talk about theoretical entities, whose existence is inferred from the percepts we perceive, and which are invoked in order to explain the order, coherence, etc., of our percepts. But since these theoretical entities can never be known some might think it preferable to postulate some other explanation, to adopt Berkeley's God rather than Locke's external objects, or even to suggest that no explanation is possible or necessary.[1]

It is often suggested[2] that Realism suffers from much the same difficulty as the Causal theory. The Realist claims to know that things exist, and what they are like, independently of our perception of them. But how can he know this if all knowledge comes, and can only come, from perception? How can perception give us knowledge of what, *ex hypothesi*, is not perceived? Like the Causal theorist the Realist is committed to a belief in facts and existences that cannot be known, while the Idealist avoids the difficulty by asserting the existence only of what is actually perceived. The position of the Phenomenalist-Idealist is harder to evaluate. On the one hand he does assert various things about what is not perceived and which therefore, according to the argument, cannot be known. But on the other hand he tries to cash these assertions in terms of

[1] Cf. Ayer IV, pp. 146-50.   [2] Perhaps the best example is Stace I.

## THE DEFENCE OF REALISM

what can be perceived, and therefore can be known. Even so, in so far as he claims knowledge about what is not perceived (even though it can be perceived) it seems that the argument will count against him also.

This does not mean that the epistemological considerations are unambiguously on the side of the strict Idealist. The claim that we perceive and therefore can know nothing but percepts involves us in what is called the Egocentric Predicament. If I can know only what I perceive, and if I perceive nothing but my own private, mental, sense-dependent percepts, then I can know nothing about anything except these private, mental, sense-dependent percepts, these contents of my own consciousness. So a strict Idealism leads inevitably to Solipsism. It is not for nothing that Idealists usually pin their hopes to Phenomenalism. No doubt a sufficiently hardheaded Idealist will welcome his fate with open arms, saying 'This is just what I have always maintained, this is all that we can really know', but this conclusion hardly makes Idealism attractive or acceptable. Idealism only avoids the Realist's epistemological problem by changing 'How can we know?' to 'We can't possibly know'. The medicine seems more fatal than the disease.

Nor is this all. I do not think that the Idealist's epistemological argument is successful in the first place. The claim is that we cannot know about what we do not perceive. We shall see (10.4) that the important conditions for knowledge are (1) that what is said to be known be true, and (2) that the person said to know have the right to be sure that it is true. For the Idealist's argument to succeed one or both of these conditions must fail to be satisfied. Naturally the Idealist will say that when I claim to know, e.g. that this table continues to exist even when I do not perceive it, the first of these conditions is not satisfied, i.e. that what I claim to know is false. But although this will be a reason for his denying that I know it, he cannot use it in an argument designed to show or suggest that the table does not exist unperceived, for that would be begging the question. If the Idealist's argument is to succeed it cannot rest simply on the claim that what is in dispute is false. It must rest on the claim that we do not have the right to be sure that it is true. And this claim, I think, abuses our ordinary concept of Knowledge. There are certain conditions for having the right to be sure that have to be satisfied before a person can correctly be said to know something. What these conditions are, in the case of our ordinary concept of Knowledge, will be determined by the circumstances in which we are ordinarily prepared to assert of someone that he knows something, assuming that that something is true. The question is whether these accepted conditions, the conditions encapsulated in our ordinary application of the word 'know', are satisfied in the case of our alleged knowledge that things continue to exist unperceived. There can be little doubt that they are. We all ordinarily say and believe that objects do exist unperceived, and even if this is a mistake on our part the fact remains that we do accept that we have the right to be sure of it (we shall see, 10.3, that the fact that a man is mistaken about something does not mean that he cannot have the right to be sure of it). We undeniably do allow the evidence of our senses to be sufficient to assure us that the things we perceive continue to exist unperceived, so any claim that we do not have the right to be sure must involve some implicit interference with those conditions for having the right to be sure that are contained in our ordinary concept of Knowledge. The only thing that could show that we do not know that objects exist unperceived would be that this is, in fact, false, and we have already seen that although the Idealist will insist that it is false, he cannot base his argument on that without begging the question.

It is important not to make too much of this point. The ordinary language argument that of course we know that objects exist unperceived, and that anyone who denies this is playing fast and loose with the ordinary word 'know', establishes only that, given the conditions for having the right to be sure that are involved in our ordinary application of the word 'know', we have the right to be sure that objects exist unperceived. There is a temptation to argue that since we, undeniably, do know that objects exist unperceived, and since a thing cannot be known unless it is true, it must be the case that objects exist unperceived. This counter-argument to the Idealist is in the odd position of making what we say determine the facts—a similar argument could, at the appropriate times, have been used to prove anything from the flatness of the earth to the existence of Gods in the trees. The fact that we all ordinarily say that objects exist unperceived does not prove that they do. What it does prove is that the ordinary standards of having the right to be sure are satisfied in this case. The conclusion is that neither the Idealist's argument that since we cannot know that external objects exist we should not assume their existence, nor the Realist's counter-argument that our ordinary use of language shows that we undeniably do know that they exist, are successful. There remains the question of what does give us the right to be sure that external objects exist. This is the question that we will be discussing in the second part of this book.

Despite all this I think that the argument against the Causal theorist still holds good. The point was that the Causal theory denies that we ever perceive external physical objects, such as tables and chairs. But although we all ordinarily agree that such external objects do exist, I do not think we would allow that we had the right to be sure that they do if we thought that we never perceived them. It is because we perceive tables and chairs as we do, that we agree that we know there are external objects. If it were agreed that, as the Causal theorist argues, we do not perceive such things after all, then I think we would say that this undermined any claim to know that they existed.

A final point: although the Idealist would be wrong to argue that even if external objects do exist we cannot know that they do, he might well argue that we cannot know this, given some special, stringent, perhaps philosophically useful, concept of Knowledge. Given this revised concept of Knowledge, he might say, we do not know, have the right to be sure of, something unless we actually perceive it, and this concept of Knowledge, though perhaps tiresome in practice, is philosophically preferable in that it allows us to take nothing for granted. But the question is whether its adoption is philosophically desirable if it commits us to Solipsism.

### 8.2 MOORE'S PROOF OF AN EXTERNAL WORLD

This brings us to other attempts to prove Realism by reference to common sense or ordinary language. The classic statement of this appeal to what we all ordinarily think and say is to be found in Moore's *Proof of an External World*. The argument here is much more than the simple-minded 'We say it so it must be right'. It runs something like this:

(1) I am holding up a human hand.
(2) Human hands are physical objects.
(3) Physical objects are external objects.
(4) Therefore I am holding up an external object.

(5) Therefore an external object exists.

This argument is directed against the Idealist but it can be reformulated as an argument against the Causal theory (1 becomes 'I am perceiving a human hand' and the conclusion is that we can perceive external objects).

The natural line to take seems to be that (2) and (3) are analytic, and that (1) is, in some sense we shall have to consider, obvious and unquestionable. Now I have already suggested (6.2) that (3) is not true by definition,[1] but whether or not the suggestion is accepted the fact remains that (3) is analytic only if 'physical object' has its ordinary, Realist, sense. It is quite open to the Idealist to give 'physical object' some other, if related, sense in which it is not even true, let alone necessarily true, that physical objects are external objects. Berkeley, for example, was quite prepared to talk about physical objects, but not to talk about external objects, from which it follows that what he meant by a physical object was not precisely what we ordinarily mean by that expression. So the argument succeeds only if the first step establishes that what is being held up is a hand which is a physical object in the ordinary, Realist, sense.

This first step might be supported in either of two ways.[2] One is by means of a paradigm case argument, to the effect that this is a paradigm example of a human hand, such that to deny that it is a human hand is to cut this expression off from the very cases from which it gets its meaning. If this isn't a human hand what can be? Now the paradigm case argument works only where the word in question can be defined solely in terms of the observable and ostensively indicable features of the object by reference to which that word is defined. In one way 'hand' is such a word, for to be a hand is just to possess certain observable and ostensively indicable features, but if the argument is to follow it must be shown that it is a hand in a stronger sense than this. It must be shown that it is a hand in a sense in which it follows that it exists even when not perceived. And since that this is so is something that goes beyond

[1] Moore's argument (VII, p. 144) that (3) is true by definition rather misses the point. He says that it is part of the meaning of e.g. 'real soapbubble' that the item be capable of existing unperceived. Certainly this is part of the meaning of '*real* soapbubble', but the question is whether it is part of the meaning of 'soapbubble' *tout court*.

[2] It does not matter for our purposes which of these methods Moore himself intended to use.

the observable and ostensively indicable features of the object, the paradigm case argument cannot establish that it is a hand in this stronger sense.

The second way of supporting the first premiss is to argue that we cannot show it to be false, because it is so obviously true as always to be more certain that the premisses of any argument designed to show that it is false. When faced with a choice between 'This is a hand' and the premisses of any argument designed to show that this is not a hand we will always choose the former. Now certainly no philosopher has ever been so rash as to deny that this sort of thing is a hand, but, as Moore himself allows, they can disagree about what human hands are. The question is not so much whether it is undeniable that this is a hand as whether it is undeniable that this, a hand, is a physical object in the ordinary Realist sense. This, clearly, is not undeniable, and to insist that it is is really nothing more than a refusal to argue the point at issue between Realist and Idealist. So although it is undeniable that this is, in some sense, a hand, it is not undeniable in any sense from which it follows that this object exists unperceived. And what is in dispute is not whether human hands exist but whether they really exist, whether they are objects that continue to exist unperceived.

In short the Idealist has Moore, and those who argue like him, in a dilemma. Either the first premiss is not strong enough for the third premiss to follow or, if it is strong enough, its truth cannot be established in the way Moore tries to establish it. To some extent Moore is aware of this when he allows that the statement 'This is a hand' can admit of different analyses, but he does not seem to see that it is only for one of these analyses that his argument holds. And, of course, whether this analysis is the correct one is precisely what is at issue in the first place.

## 8.3 THE ARGUMENT FROM THE CONCEPTUAL SCHEME

Ordinary language is inescapably Realist. This fact might be used to give a more substantial support to the 'We all say it so it must be so' argument. For it can be argued that 'Everyone agrees that there are external objects' differs from 'Everyone agrees that the world is flat' in that the assumption of the existence of external objects is absolutely crucial to our entire way of thinking and talking, and is therefore not open to revision or rejection in the way that the belief that the world is flat is. In biting at external objects the Idealist finds himself chewing our entire conceptual system, and this is altogether too much to swallow. Certainly the denial of external objects has a far more widespread effect on our ordinary opinions than some philosophers, Berkeley for example, may have noticed, but it is not clear that this raises any new or insuperable difficulties for the Idealist. Nor is it clear whether a similar argument can be brought against the Causal theory. The Causal theory might well be regarded not as in opposition to our fundamental ways of thinking about the world, but rather as an extension and explanation of them.

The suggestion is, then, that the assertion of the existence of external objects is absolutely fundamental to our conceptual scheme,[1] such that to deny this is to reject our conceptual scheme. The argument seems to be first that some method of identification and reidentification is necessary for language, truth and falsity, any conceptual scheme, in the first place, and second that our method of identification and reidentification presupposes the existence of external, and in particular physical, objects. Given this it follows that it is only through the assumption of the existence of external objects that our conceptual scheme gets going in the way that it does. But this does not show that this assumption is an essential presupposition of our present conceptual scheme, so long as it is possible for this same conceptual scheme to get going in some other way. The question is: how different could our methods of identification and reidentification be before the conceptual scheme would cease to count as this, our present conceptual scheme. Or: what individuates conceptual schemes, distinguishes one from another? The answer may well be that it is essential to this conceptual scheme that we identify and reidentify in terms of external physical objects located in a spatio-temporal framework, and that any conceptual scheme which did not base its identification and reidentification of particulars upon this framework and the external physical objects located in it could not count as this conceptual scheme. But I am not sure how this can be proved.

So far, then, let us say that Idealism involves not only the rejection of certain fundamental beliefs about the nature of the world we perceive, but even the rejection of the conceptual scheme in terms of which we think and talk about that world. This makes the

[1] Cf. Strawson I and Hampshire III.

choice between Realism and Idealism more than a choice between philosophical theories of perception; it makes them a choice between rival conceptual schemes. Now this does not make Idealism any less acceptable, unless we can also argue that an Idealist conceptual scheme is inadequate or even impossible. It might, for example, be argued that an Idealist conceptual scheme makes it impossible to state various things which can be stated in our ordinary Realist one. This is a very strong claim which cannot, perhaps, be settled without going to the lengths of developing an Idealist conceptual scheme, but prima facie it is not very plausible. The mere fact that he denies the existence of external objects, for example, does not mean that the Idealist cannot have the concept of External Object, any more than the fact that we deny the existence of unicorns means that we cannot have the concept of Unicorn. The truth must be quite the opposite; we could not deny their existence if we did not possess the concept. Of course if it can be shown that we cannot have the concept of External Object unless we assume the existence of such things it would then be shown that an Idealist conceptual scheme must be inadequate. But once again I do not see how this can be shown.

It seems to me that those who argue[1] that an Idealist conceptual scheme is impossible or incoherent tend to slip from the claim that the existence of external objects is a necessary presupposition of our conceptual scheme to the claim that it is a necessary presupposition of any conceptual scheme. Identification and reidentification are necessary for any conceptual scheme, but it is far from obvious that identification and reidentification are possible only given the existence of external objects. No doubt, as Strawson argues, when with our present conceptual scheme we identify something as 'one and the same again' we imply—perhaps this is even an entailment, although I find that doubtful—that the item in question has existed even through the time when we did not perceive it. But I see no reason for maintaining that this implica-

---

[1] Cf. Hampshire III, p. 17, p. 36, although I cannot extrapolate the arguments that are supposed to show that 'we must unavoidably think of reality as consisting of persisting things'. I am not sure what Strawson's position is on this. He does say (I, p. 35) that scepticism over the existence of external objects is incoherent, but in the same passage he says we may regard this as a sketch of an alternative conceptual scheme. These two accounts of the sceptic's position seem to me to be incompatible, but in so far as Strawson accepts the latter account he cannot be rejecting an Idealist conceptual scheme as incoherent.

tion—or entailment—must hold for all possible conceptual schemes. Certainly Strawson tends to talk as though he has shown that reidentification is impossible without assuming that items continue to exist unperceived, but I have argued elsewhere[1] that the most that Strawson can legitimately claim to have shown is that this is an essential presupposition of our present conceptual scheme.

What may be the case is that the only conceptual scheme which we mere humans could, as a matter of empirical psychological fact, develop, *ab initio*, is a Realist one, one that presupposes the existence of external objects. It is a familiar point that our conceptual scheme is established and communicated by our language, and that our language is, and can only be, taught and learnt and carried on via reference to external objects. This means that without the assumption of external objects we would not, and could not, acquire any language or conceptual scheme at all. To this might be added the suggestion that the Realist interpretation of our experience is, somehow, instinctive, perhaps for Darwinian, survival-of-the-fittest, reasons. Yet all this is a long way from showing that a Realist conceptual scheme is the only possible, or even the philosophically preferable conceptual scheme. Quite apart from the fact that all this is at best a matter of empirical fact, and the fact that the Idealist may well agree that the assumption of external objects is necessary in this way while adding that, necessary as it is, it is a mistake all the same, there is the fact that nothing has yet been said to show that it is impossible to develop a different conceptual scheme, one which does not assume external objects, after we have first acquired this Realist one.

The conclusion is, I think, that the choice between Realism and Idealism is, in the end, a choice between radically different interpretations of our experience, and hence between different conceptual schemes in terms of which we describe and account for that experience. This is much the same point as that made by Ayer when he talked of the different theories as 'alternative languages'. And to the problem of the final choice between these rival interpretations of our experience we must now turn.

[1] I

# CHAPTER 9
# THE CHOICE BETWEEN THE THEORIES

## 9.1 THE THEORIES CONSIDERED

We have now discussed the various arguments that are usually brought forward in support of one or other of the three traditional theories of perception, and I think we must conclude that none of these arguments succeeds in proving any one theory as against the others. The choice must be made on theoretical, methodological or heuristic grounds; if we cannot prove that one rather than another is the correct theory then we must decide, by reference to simplicity, convenience and the like, which one to regard as correct. The choice between Realism and Idealism, in particular, seems, in a perfectly familiar sense, to be a matter of metaphysics, both in that it involves a decision about something which is not capable of empirical proof or disproof, and in that it seems to involve a decision between rival conceptual schemes, rival ways of interpreting and describing our experience.

In deciding between the theories we must first consider them from the point of view of ordinary language and common sense. Clearly Realism is the preferable theory on this score, for both common sense and ordinary language are essentially Realist. Adoption of a Causal theory would involve a certain amount of tampering with our common sense beliefs, and Idealism runs directly counter to our accepted views on and talk about the world. Phenomenalism hopes to provide a means of making Idealism conform to common sense, an interpretation or analysis of ordinary language that will make it consistent with Idealism, but we have seen that Phenomenalism, particularly in the strict form the Idealist needs, fails. From this point of view, then, Realism is the theory to adopt. Someone will insist that there is nothing sacrosanct about common sense and ordinary language—common sense might be mistaken and ordinary language might be based on incorrect assumptions. And so they might. The point is not that if it is in accordance with common sense and ordinary language it must be correct, but that if it is in accordance with common sense and ordinary language it must be familiar, we must already know and understand its terminology and implications. From a purely heuristic point of view the familiar is, *ipso facto*, preferable to the unfamiliar. There is the additional suggestion that if the Realist interpretation of our experience were not the simplest and best for any number of reasons it would not have become that of ordinary language and common sense in the first place. It can be no accident that ordinary language is Realist—this may even be a consequence of the nature of language itself—and it seems almost a psychological impossibility to convince oneself of the truth of Idealism. Perhaps, as suggested earlier, the assumption of the existence of external objects is inevitable for Darwinian, survival-of-the-fittest, reasons. The argument is, in effect, that if Realism has been the universally accepted theory to date there will have to be very good reasons indeed for dropping it now.

The second consideration is that of Occam's razor, and this seems to favour Idealism as much as the first consideration favours Realism. The reduction of all existence to two types of things— 'spirits' and 'ideas'—is one of the most striking features of Berkeley's philosophy, and Neutral Monism provides the greatest simplicity possible when it tries to reduce both minds and bodies to percepts. However Occam's principle is that entities must not be multiplied *beyond necessity*, and it might be held that it is preferable, in a way necessary, to postulate more than the fewest possible kinds of entities. As a reduction to a minimum Idealism, and Neutral Monism even more, are striking, but their success is achieved at a price—Solipsism—which we may prefer not to pay. The Causal Theory, on the other hand, seems, from the point of view of Occam's razor, to be in an even worse position than Realism. Both postulate two kinds of entity—external objects and percepts—where Idealism postulates only one, but the Causal theorist is, in addition, committed to a thorough-going duplication of percepts and external objects. The Realist may prefer external objects to Solipsism, but this can hardly justify the Causal theorist's massive duplication of external objects and percepts.

Next, epistemological considerations. At first sight these too seem to favour the Idealist, but we have seen that the Idealist's scruples about what can be known commit him, in the end, to some form of Solipsism. Even so the Idealist does make use of stricter standards of knowledge than does the Realist and this, for a philosopher, might count as an advantage. The question is whether we prefer the difficulties to the strictness. Epistemological considerations count strongly against the Causal theorist. We have seen that for him external objects must have the status of theoretical entities whose existence is inferred in order to account for and explain what we perceive. This unverifiable inference will be the only way in which we can know about external objects.

Two other points arise from our discussion in the preceding chapters. First, the facts of physics and physiology are alleged to support the Causal Theory, although I have argued that this is so only given a special and I think implausible interpretation of those facts, i.e. the correlation of the brain activity involved in perception not with perceiving itself but rather with what is perceived. Second, the occurrence of hallucinations and their similarity, or at any rate possible similarity, with the objects of veridical perception is cited in favour of non-Realist theories, on the principle that if things are qualitatively alike they are unlikely to be of radically different ontological types. The question is whether this principle is sufficient to outweigh the disadvantages of the theories it commits us to, particularly in view of the fact that hallucination is less often like veridical perception, and less common, than those putting forward the argument usually suggest. This appeal to hallucination, as we have seen, has gained much of its support from the failure to distinguish hallucination from illusion and other types of non-veridical perception.

There is one last point, concerning the explanation of what we perceive. The fact that what different people perceive via their different senses and at different times fits in consistently, for the most part, from person to person, sense to sense, and time to time, is always cited in the attempt to justify, or at any rate explain, our belief in the existence of an external world. How is this consistency, this constancy and coherence in what we perceive to be explained? Realist and Causal theorist both, in their different ways, explain it by reference to external objects, but the Idealist, unless he is prepared to follow Berkeley in basing everything in God, has no explanation to offer. We might turn Berkeley's objection[1] against Locke on its head, and protest that if there were external bodies our experience would have to be as it now is, and if there were not it would be inexplicable how experience comes to be as it is.

To sum up: Realism is supported by the appeal to common sense and ordinary language, i.e. to the preferability and simplicity

[1] I, § 20.

of sticking to accepted and understood ways of thinking and speaking, and by the point about explanation. Idealism is supported by the appeal to Occam's razor and, to a limited extent, the point about hallucination. But on the other hand it conflicts directly with common sense and ordinary language and seems unable, except by some appeal to a *deus ex machina* to explain why what we perceive is as it is. Above all it commits us to Solipsism, for which its use of a specially strict concept of Knowledge seems poor recompense. Finally, the Causal Theory seems to have little to support it except the point about hallucination and a dubious interpretation of the facts of physics and physiology, and is at a considerable disadvantage from the epistemological point of view and as regards Occam's razor, as well as, to a certain extent, running counter to common sense.

The conclusion seems to be that the preferable theory is the Realism which we all ordinarily and unquestioningly assume. Indeed it is for this very reason that it is the preferable theory. Not because it must be true if we all say it but because if we all say it, and no proof that it is false is forthcoming, it is best to accept it as true, rather than adopt some unfamiliar and difficult alternative. Better the devil you know, and can handle, than the devil you don't know. So my choice is for Realism and in the second part of this book when I consider our knowledge of the external world, I shall take this Realism for granted. That is I shall take it for granted that there are external objects, objects existing independently of our perception of them and that we can and do perceive them. But I do not claim to have refuted either Idealism or the Causal theory. So far as I can see it remains possible, though neither plausible nor tempting, to interpret our perception as perception of percepts caused by the action of external objects, or even to deny the existence of such external objects altogether.

## 9.2 THE SENSIBILIA THEORY

I have left till last a fourth theory of perception, a rather strange combination of all three theories which I will call the 'Sensibilia' theory. I have left it till last because it seems to me by far the most implausible, involving most of the difficulties faced by each of the three other theories.

The crucial difference between sensibilia and percept theories of perception is that although the former, like the latter, hold that we perceive private entities which are not at all like the objects we normally think of ourselves as perceiving, the sensibilia theory continues to regard these private entities as physical, i.e. located in physical space, and external, non-sense-dependent. This theory has been adopted by Russell, in particular, although Broad's theory of sensa seems a close relative and some other philosophers[1] have not clearly distinguished sensibilia theories from percept, or as they may prefer to call them 'sense data', theories of perception. But since Russell has been the only one to develop an explicit sensibilia theory it is his formulation that I shall discuss.

The fundamental notion in Russell's theory is that of a 'perspective', which is what we might ordinarily describe as the appearance the world presents from a certain point of view. It is not clear whether these perspectives are private. Russell seems to think that they must be because two people cannot perceive from the same place[2] but he does not consider the possibility of one person's taking another's place before the things perceived change. And since he allows that perspectives can exist unperceived—for not all points of view are occupied—it seems that it should be possible for different people to perceive the same perspective at different times. A perspective includes, indeed consists of, sensibilia, i.e. of the appearance of the various objects, sounds, etc., which can be perceived from the point of view in question. A perceived sensibile is called a sense datum, and Russell allows that sensibilia may be altered in the act of being perceived.

A perspective is three-dimensional, for the world appears three-dimensional, and it is also located in three-dimensional space. So Russell speaks of the world of prespectives as six-dimensional, consisting of three-dimensional items themselves located in three-dimensional space, the point being, presumably, that the three-dimensional perspectives are extended in different dimensions from those they are located in. This world of perspectives, our real world, consists of an infinite number of these perspectives, and physical objects, sounds, etc., are constructed from perspectives, or more accurately from the sensibilia which figure in perspectives. The object is the class of its appearances, the sum total of all the sensibilia located at, i.e. to be perceived from, all the different points of view from which the object can be perceived.

When fully worked out all this talk about external but apparently

---
[1] Cf. Price I, Hirst I, ch. 3.     [2] Cf. II, p. 138.

private appearances of objects might be construed as merely an alternative reformulation, a re-statement, of the ordinarily accepted Realist interpretation of perception and the external world. But for Russell it is more than that. Physical objects are, for him, logical constructions from sensibilia, and as such they are metaphysical fictions to be eliminated in favour of sensibilia. In short, sensibilia and the perspectives they make up are the only external objects.

Now if this construction, and elimination, is to work we should begin with sensibilia and work up to physical objects and the ordinary Realist description of the world. This Russell fails to do; his definition of the material thing as the class of *its* appearances is obviously circular, and any attempt to specify the appearances without referring to the object will run into the same difficulties as those faced by the Phenomenalist.[1] The main charge Russell must face is that of reifying appearances. Sensibilia are defined as the appearances things present, and are the assigned a location, or at any rate one location, quite separate from that of which the are appearances, as if appearance and object were, or could be, separate items. One apparent souce of this mistake is the familiar appeal to the fact that things can appear to have different qualities from different points of view. Russell insists[2] that the object cannot be both square and oblong, and avoids this 'difficulty' by suggesting that these different qualities are not in the same but in different places, viz. the different places from which those shapes are perceived. But even this does not avoid the alleged difficulty, because people can perceive different qualities from the same place. True, Russell may want to say that these different appearances appear in private worlds, private places, and so are still not in the same place, but that would conflict with his other claim that the perspectives which such appearances make up are located in public space. However the 'difficulty' disappears the moment we refuse to regard the appearance as a thing in its own right, and recognize the obvious fact that although a thing cannot *be* both square and oblong it can *appear* both square and oblong to different people, or from different points of view. As always a simple confusion between 'is' and 'appears' lies behind the reification of appearances and leads, in the end, to a theory such as Russell's.

---

[1] Prichard I argues very convincingly that Russell's account presupposes, and so cannot replace, a common sense Realism.   [2] III, p. 153.

## THE CHOICE BETWEEN THE THEORIES

Even apart from these difficulties I doubt whether the sensibilia theory is to be preferred to the others we have discussed. Like Idealism the theory runs counter to common sense. Like Realism and the Causal theory it asserts the existence of what is not perceived—something which cannot be known on the Idealist's strict standards of knowledge—and yet at the same time the insistence on the privacy of 'sense data' (perceived sensibilia) seems to commit us to Solipsism. Like the Causal Theory it argues for the existence and nature of sensibilia from the facts about material objects ('the appearance of a thing in a given perspective is a function of the matter composing the thing and of the intervening matter'[1]) when, quite apart from the claim that material objects are fictions, the argument should run in the opposite direction. And although the sensibilia theory may share the ontological simplicity of Idealism, postulating but one type of entity where Realism and the Causal theory postulate two, it commits us to the existence of an enormous, in fact infinite, number of such entities, for there is at least one sensibile at each point in space.[2]

[1] III, p. 165.
[2] Presumably there will be different sensibilia in the one place, varying in the direction we look from that place, and if we want to say that hallucinatory sensibilia too are located in the places we perceive them we seem to be committed to the existence of an infinite number of sensibilia (all possible hallucinations) at every point in space!

# PART TWO
# OUR KNOWLEDGE OF THE EXTERNAL WORLD

CHAPTER 10
# PERCEPTION AND KNOWLEDGE

## 10.1 THE PROBLEM

There are many questions about our knowledge of the external world: how do we know what external objects are really like? how do we know what physical objects, for example, are? how do we know about objects we are not perceiving? how do we know about objects we never perceive? how do we know about objects we cannot perceive? how do we know laws and generalizations that hold true of sets of external objects, etc., etc. There is also the question of how we acquire the concepts in terms of which we think and speak about this external world. But the basic question, *the* problem of our knowledge of the external world, is the question of how we know that there exists an external world in the first place, how we know that there are objects which exist independently of our perception. And, given a Realist theory of perception, this problem becomes: how do we know that the objects we perceive continue to exist when we are not perceiving them?

The most obvious and natural answer to this question is that of the Empiricist, that it is from our perception and our perception alone that we can and do know this. But although this answer is very plausible it is not particularly clear. The problem in any philosophical discussion of perception is: how does our perception provide us with this knowledge that the things we perceive are external objects? It is with this very simple, but I think absolutely basic, question that we are now concerned.

In asking this question I am, of course, taking two things for granted: that we do perceive external objects, and that we do know that we do. I have argued for both of these facts in the first part of this book, although I now want to say a little more about the second. And the best place to begin is by asking: what precisely is knowledge?

## 10.2 KNOWING AND SAYING 'I KNOW'

There is little point in repeating familiar arguments against the view that knowledge is some form of mental state of process. Instead let us begin with what seems to be the accepted contemporary view. The details of the analysis differ from philosopher to philosopher, but is it usually something of this pattern:

Knowing that $S$ is $P$ involves:
(1) its being the case that $S$ is $P$;
(2) being sure that $S$ is $P$;
(3) having good evidence (or something of the sort) that $S$ is $P$.

Disagreement usually concerns the precise formulation of the third condition, but it seems to me that the second is equally suspect. Perhaps the best way of indicating my suspicions is through a discussion of what might be called the 'quasi-performative' theory of knowledge.

This derives from Austin's paper on 'Other Minds' where, among other things, he compares the phrase 'I know' with the phrase 'I promise', the latter being an example of what he elsewhere calls a performative. It seems that a performative is any phrase where to utter the phrase is not to describe what you are doing but to do it,[1] as to say 'I promise' is to promise and to say 'I do' during a wedding ceremony is to accept the person beside you as your lawful wedded spouse. Now it is obvious that 'I know' is not a performative, at least not in this sense, for to say 'I know' is no more to know than to know is to say 'I know'. The most we might say, as Austin does, is that to say 'I know' is, in some sense, to give one's backing for the truth of the statement in question.[2] However people are sometimes misled into speaking of knowing or knowledge as performative, or at least quasi-performative. This is clearly mistaken, and Austin himself would have been the first to insist that knowing is not in any sense a performance. It is *saying* 'I know' which is the performance, and that, as we have just seen, is quite different from knowing. This simple confusion between *knowing* and *saying 'I know'* is in fact quite common. We often hear, for example, the question 'When is knowledge justified?', a question which hardly makes sense unless we are, perhaps, worried by people with information they should not have

---

[1] Another suggestion is that a performative is any phrase where one does something, not necessarily the thing mentioned in the phrase, in uttering that phrase. This sense of 'performative' seems to be so all-embracing as to be vacuous, as Austin's own later work suggests.

[2] Perhaps this is true, but I do not see that it tells us anything about what knowledge is, or that it raises or solves any problems about what it is to say 'I know', or even that it does anything to distinguish saying 'I know $p$' from saying '$p$'. In short I do not see why this point has been thought so important.

had access to. We do not ask 'When is truth justified?'; the relevant questions are 'When is a statement true?' and 'When are we justified in saying that something is true?'. Similarly we should ask not 'When is knowledge justified?' but 'When do we know something?' and 'When are we justified in saying that we know something?'

This latter is an interesting question, for the fact is that I can be justified in saying 'I know' even when I am mistaken in saying this, even when I do not know. When I last looked around I saw a painting on the wall behind me, and I am sure that I would have noticed if anyone had come in and taken it away. So I am quite justified in saying 'I know there is a painting on the wall behind me'. And I would still be justified in saying this even if it had been removed in some ingenious way without my noticing. I would be mistaken in what I said, but I would still be justified in saying it. $P$ must be true if I am to know that $p$, but it need not be true for me to be justified in saying that I know it. What, then, is necessary for my being justified in saying that I know $p$? It seems that I must be sure that $p$ is true and have evidence that it is. In other words the last two conditions generally accepted for knowledge seem to be conditions for being justified in claiming to know.[1]

Now if I can be justified in saying 'I know' even where, as it turns out, I do not know, can I know something without being justified in saying 'I know'? Suppose that a contestant in a quiz is asked 'Who was the President of the United States during the Second World War?', and he answers, slowly and doubtfully, more as a question than an answer, 'President Roosevelt?'. Unless there is some evidence that this is a guess we would naturally conclude that the contestant did know the answer, did know that Roosevelt was President during the Second World War. But he certainly wasn't sure of it, and he would scarcely have been justified in saying that he knew that Roosevelt was the President in question. Similarly a candidate may be asked the same question and although he beats his brow and maintains that the answer is

---

[1] There is a sense in which one is justified in saying one knows even when one is convinced that the thing in question is false, as when the officer tells the trapped soldiers that he knows a way out. But this is a different sort of justification. The man who says 'I know' when he isn't sure is not just telling a falsehood, he is misusing the word. When I say I know there is a picture behind me, and it has been secretly removed, I am not misusing the word even though what I say is false.

on the tip of his tongue be unable to give it. Yet the moment he is told he exclaims 'Of course! Of course! I knew it all the time'. Indeed even if he looked quite blank when the question was put to him we might still agree that he did know it all the time so long as he instantly recognized the answer when he is told. Once again the contestant is said to know something which he is not sure of, and which he is not justified in claiming to know.

These are two cases where we say that a man knows something, in a perfectly ordinary and acceptable use of 'know', even though he is not sure of it. Being sure cannot be a necessary condition for knowledge, in at least one ordinary use of the word. Being sure is a necessary condition for being justified in claiming to know, but only those who ask such questions as 'Is knowledge justified?', and so confuse knowing with saying that one knows, will make the mistake of taking this to show that being sure is a necessary condition for knowledge. Our two examples also count against the third suggested condition for knowledge, at least so long as this is stated as that we must have good evidence, or a proof, or something of the sort, for what is known. There are other counter-examples too: I have neither evidence nor proof that Napoleon died on St Helena, but I know it just the same. No doubt there must be evidence or proof somewhere, or else we would not accept it as true, but that is a different matter.

### 10.3 THE RIGHT TO BE SURE

Nevertheless some form of the third condition does seem to be necessary for the analysis of knowledge, and I think the best formulation uses Ayer's phrase 'the right to be sure'.[1] One can have the right to be sure without actually being sure, and without being able to adduce evidence for or a proof of whatever it is that one has the right to be sure of. It is because I have the right to be sure—because someone told me, because I read it, or some such thing—that Napoleon died on St Helena that I can be said to know that he did. I don't have to remember where or how I learnt this fact—a person can know without knowing how he knows, without knowing what gave him the right to be sure. It is even possible, as in our quiz examples, for a person to know without knowing that he knows. In fact this must be so if we are to avoid infinite regress (Do I have the right to be sure that I have

[1] Cf. VI, ch. 1.

the right to be sure that I have the right to be sure . . . ?). This formulation of the third condition for knowledge also has the advantage of enabling us to talk without oddity of cases where it would be odd to talk about proof or evidence. I do not have proof or evidence that I like strawberries but I do have the right to be sure of it, and so I can be said to know that I like strawberries. Notice, finally and importantly, that we can have the right to be sure of something even though it is false. In our earlier example I have the right to be sure that there is a picture on the wall behind me, even if someone has been clever enough to remove it without my noticing. Or if a normally reliable authority tells me that Napoleon died on Corsica I have the right to be sure of this even thought the authority is, for once, mistaken.

There is, however, one important objection to the use of this phrase, 'the right to be sure' in the analysis of knowledge.[1] It seems to be in as much need of explanation as was the original term 'knowledge'. It is, after all, a technical phrase and using it will do nothing to elucidate knowledge until it is itself elucidated, in particular until we are told just when it is that we have the right to be sure of various things. However to do that would be to deal with the entire subject matter of epistemology, for the epistemological question 'How do we know?' is the question 'What is sufficient and necessary to give us the right to be sure?' The phrase 'the right to be sure' is a handy title which enables us to isolate the epistemologically important element in knowledge. That it is the important element is shown by the fact that in our two quiz examples we allow that the contestants know that Roosevelt was the President in question only in so far as we allow that, although they were not sure, they had the right to be sure and were not merely guessing or pretending. Again if we have doubts about

[1] Strawson II objects to the phrase because it suggests that we ought not be sure unless we have the right to be sure, when 'it is not clear that the degree of one's conviction is a matter of will' (p. 305). But we might compare the right to be sure with, say, the right to lose one's temper. There are circumstances in which one is quite justified in losing one's temper, even though it is not clear that losing one's temper is a matter of will. And even apart from this it does not follow from the fact that one does not have the right to be sure that one ought not to be sure. A hitch-hiker may not have a right to the empty seat in the car, but that does not mean he ought not sit there. I would also reject Strawson's suggestion that 'a man must have the right to say he knows, if he knows'. Perhaps the word 'right' is misleading—we are not speaking of a legal, moral or human right. To say 'He has the right to be sure' amounts to saying 'He would be justified in believing (whether he does or not)'.

whether something can be known these are usually tantamount to doubts as to whether we can have the right to be sure of such things. A person who forecasted the future correctly time after time by examining the leaves in a teacup would be said to know what is going to happen only in so far as we would be prepared to say that his unusual method gave him the right to be sure that those things were going to happen. The debate as to whether he knows the future is a debate as to whether his evidence is sufficient to give him the right to be sure.

So in discussing our knowledge of the external world I shall be concentrating on our right to be sure of the things we are said to know. Epistemologically speaking, this is the important thing.

### 10.4 AN ANALYSIS OF KNOWLEDGE

Knowledge involves more than truth and the right to be sure. If I was told as a child that my great great grandfather was an Irishman but have long since forgotten this I cannot be said to know it, nor to have the right to be sure in any sense in which the right to be sure, plus truth, is sufficient for knowledge. We need some further condition for knowledge which will clarify the notion of having the right to be sure by showing when the right to be sure gives us knowledge.

I think this third condition will be a variable one. In our second quiz example we said that there was a sense in which the contestant knew the answer to the question. But there is also a sense in which he did not know the answer, or else he should have been given the prize. It is natural to speak of types or levels of knowledge here, with the 'level' of knowledge depending on which form of the third condition is satisfied. At the lowest level of knowledge, in the sense of 'know' in which the contestant did know the answer, what is necessary is that the person be able to recognize the thing in question as true, even though he may be unable to offer it as the truth until someone reminds him (I do not know that my great great grandfather was an Irishman because I would not recognize this as true if someone told me). At the next level, in the sense of 'know' where the contestant did not know the answer, what is necessary is that the person have the right to be sure and be able to offer the truth as the truth, even if he is not sure of it. At a higher level knowledge involves being sure that it is the truth, and at a higher level again it involves being able to offer evidence or proof that it is true. And finally we have the highest level where knowledge involves being absolutely certain, in the sense that the person cannot be mistaken about what he knows.[1] Notice that although being sure is not a necessary condition at all levels of knowledge it is at least necessary that we be not convinced that the thing in question is false. For if we were we would not even accept it as true.

There is yet another condition for knowledge. Examples have recently been given[2] where a person is sure, and has the right to be sure, of something that is true, and yet cannot be said to know it. The butcher tells me that the price of meat has come down. He is an authority on the price of meat so I have the right to be sure, and in fact prices have come down. But the butcher thought they had gone up, and had lied to me in order to attract me into his shop. In some way my 'knowledge' is based on false grounds, and hence does not really count as knowledge, even though I am sure, have the right to be sure, and it is true. I think the fourth condition needed to rule such cases out is something like this:[3] what gives me the right to be sure must be something which would not be but for the truth of what is known (it may even be that which is known itself). Thus I know that I have a pain because what gives me the right to be sure is the pain itself. If there were no pain there would be no such thing to give me the right to be sure. Similarly if the teacher tells me that Napoleon was born on Corsica this gives me the knowledge that Napoleon was born on Corsica only in so far as it is true that the teacher would not have told me this

---

[1] Just as some philosophers have insisted that we do not know something unless we cannot be mistaken about it, so other philosophers have turned this on its head and claimed that we cannot be said to know something unless we can be mistaken about it. The reason for this paradoxical view seems to be that if there is no possibility of mistake there is no point in saying 'I know'. I don't see how this shows that we are wrong to say 'I know' in such circumstances. Nor is it hard to think of circumstances where there is a point to saying 'I know' even though there is no possibility of mistake, e.g. when we are discussing the things we know and the different ways in which we know them. Sometimes we are told that we should, in such cases, speak not of knowing, but only of being able to say! But there are many things we can say which are false and even nonsensical, and if this phrase 'can say' is to be explained in a sense that rules out the false and the nonsensical, it seems that it will have been given a sense synonymous with 'know' after all. So far as I can see this is a quite gratuitous and pointless terminological innovation. Certainly the ordinary man sees nothing strange about the claim that we know when we are in pain and when we are not.
[2] Cf. Gettier I, Clark I.
[3] I believe I owe this suggestion to Max Deutscher.

if Napoleon had not been born there. Notice that this is a condition for knowledge, not a condition for having the right to be sure. Even if the teacher had been mistaken, or lying, or would have told me this regardless of where Naploeon was born, I would still have the *right to be sure* that he was born on Corsica. But I would not *know* that he was.

I suggest, then, the following analysis of knowledge:
I know that $S$ is $P$ if:
(1) It is true that $S$ is $P$;
(2) I have the right to be sure that $S$ is $P$;
(3) I am either (a) able to recognize '$S$ is $P$' as the truth, or (b) able to offer '$S$ is $P$' as the truth, or (c) sure that $S$ is $P$, or (d) able to offer evidence or a proof that $S$ is $P$, or (e) absolutely certain that $S$ is $P$, in the sense that I cannot be mistaken about it.

(4) My right to be sure depends on the truth of '$S$ is $P$', in that if $S$ were not $P$ whatever it is that gives me the right to be sure would not have occurred.

(5) What enables me to satisfy condition (3) is something which gives me the right to be sure that $S$ is $P$.

The last condition is needed to rule out possibilities like this: I may have the right to be sure that my great great grandfather was an Irishman, because someone once told me, and a hypnotist may induce me to accept this as true when someone mentions it, without this counting as knowledge, because what leads me to accept it as true is not something that would give me the right to be sure.

## 10.5 SCEPTICISM

The sceptic maintains that we do not know whether external objects exist. There are many ways in which this might be argued.

First, it might be held that we do not know that external objects exist, because it is not the case that they do. To say this would be to adopt Idealism, and we have already considered this alternative. We have decided to adopt Realism, and so this form of scepticism does not arise for us. Nevertheless it may still seem that it does, thanks to a misinterpretation of the analysis of knowledge. We have said that 'External objects exist' is true, but we have not shown that it must be true; for all we have said, it could still be false. Now it seems that if something is known it must, according to the analysis of knowledge, be true. And this might look as though it meant that it cannot be known unless it is necessarily true. So it seems that we cannot know that external objects exist. However, what is necessary for knowledge is that the statement in question be true, not that it be necessarily true; 'Only what *is* true can be known', not 'Only what *must be* true can be known'. The mistake is to take 'If something is known it must be true' to mean 'If something is known then it is necessarily true'. What it should mean is 'If something is known then-necessarily it is true'.

The next sceptical claim is that, whether or not it is true, we do not have the right to be sure that external objects exist, and so cannot know that they do. This scepticism may take two forms. The first depends on the fact that we do not have a conclusive, let alone a logically conclusive, proof that external objects exist. Although we have decided to accept Realism we might, after all, be mistaken in this and, the argument runs, since we might be mistaken we cannot have the right to be sure that external objects do exist. The familiar reply is that it is not necessary for our having the right to be sure of something that there be no possibility of our being mistaken about it. I might be mistaken in thinking that Napoleon died on St Helena, but this does not mean that I do not have the right to be sure of it. Indeed provided that my source is reliable and reputable I have the right to be sure of it even if it is false, although then, of course, I cannot be said to know it. If the sceptic, from a desire to put everything on the level of mathematical knowledge, insists that the possibility of mistake means that we cannot have the right to be sure, this means that he is being much stricter about knowledge than we ordinarily are, is restricting knowledge to what I called its highest level, and so is, in effect, proposing a change in our ordinary concept of Knowledge. He is a revisionary 'with whom we do not wish to quarrel, but whom we do not need to follow'.[1] He doesn't succeed in proving that we don't know what we ordinarily claim to know, so long as 'know' is being used in its ordinarily accepted sense, although he does succeed in showing that we do not know in his own special sense of that verb. We might draw a distinction[2] between 'descriptive' and 'revisionary' epistemology, between the discovery and examination of our accepted standards of having the right to be sure and what is accepted as meeting them, and the suggestion and examin-

---
[1] Strawson I, p. 36.
[2] Following Strawson's distinction between descriptive and revisionary metaphysics, I, p. 9.

ation of different, probably more stringent, standards of having the right to be sure and what might be accepted as meeting them, the eventual aim being, presumably, to cast light on the nature of our knowledge.

The second argument that we do not have the right to be sure that external objects exist is that the various things we appeal to in trying to establish that we know, have the right to be sure, that they exist do not establish, logically, that they do exist.[1] Suppose, for example, that we appeal to the evidence of our senses in trying to show that we do have the right to be sure that external objects exist. The evidence of our senses could be just as it is and yet it not be true that external objects exist. So how can this evidence give us the right to be sure that they do exist? The answer to this is much the same as with the previous argument. Our ordinary concept of Knowledge, as expressed by the ordinary use of the verb 'know', encapsulates certain conditions for having the right to be sure, and what these conditions are, and what counts as satisfying them, is determined by where we are and where we are not ordinarily prepared to say of someone that he knows something, assuming that thing to be true. If in situation $x$ we are all ordinarily prepared to say of $A$ that he knows $p$, it follows that in situation $x$ the ordinary conditions for having the right to be sure of $p$ are satisfied, although, obviously enough, it does not follow from what we are ordinarily prepared to say that $p$ is true. Now we are all ordinarily prepared to say that we know that objects continue to exist even when we do not perceive them. It does not follow from this that they do exist unperceived, but it does follow that the conditions, whatever they may be, for having the right to be sure are satisfied. The sceptic is, implicitly, revising our ordinary concept of Knowledge, and setting up new conditions for having the right to be sure. Notice that this does not mean that we can never show that we do not know something which we all ordinarily claim to know. There are two ways in which a universal claim to knowledge can be shown to be mistaken: by showing that what we claim to know is in fact false, or by showing that we do not, after all, have the evidence for it that we thought we had (e.g. by showing that we do not, after all, perceive external objects). What we cannot show is that this evidence is not good evidence, not sufficient to give us the right to be sure.

[1] Cf. Ayer VI, ch. 2, § ix.

So much for the attempt to show that we do not know that external objects exist, either by claiming that this is false or by claiming that we do not have the right to be sure that it is true. The next form of scepticism is the argument that even though we may, in fact, know that external objects exist we cannot *prove* that we know it, and this, once again, may take two forms: the argument that we cannot prove that it is true, and the argument that we cannot prove that we have the right to be sure of it. Now, just as the argument that we do not know that external objects exist is often based on setting specially stringent conditions for knowledge so the argument that we cannot prove that we know it is often based on setting specially stringent standards for proof. Perhaps it is impossible to give a logically conclusive proof that external objects exist, but why should we have to give a logically conclusive proof? Surely if we have good evidence for $p$, and no evidence against $p$, then we have proved $p$, even though there remains the possibility of mistake? The sceptic may reply that although such a 'prima facie' proof may do for all practical purposes and until something better comes along, it still isn't a real proof. But here, as so often, the adjective 'real' gives the game away. Just as we say to the sceptic who says that we don't really know, that his real knowledge' is a special sort of knowledge and that what is true of it need not be true of knowledge in the ordinary sense, so we can say to the sceptic who says that we can't really prove it, that his 'real proof' is a special sort of proof, and that what is true of it need not be true of proof in the ordinary sense.

Moreover, in so far as it is possible to prove that we, people generally, have the right to be sure of something it is possible to prove that that something is true, for to have the right to be sure of $p$ is to have the right to be sure that $p$ is true, and so to prove that we have the right to be sure of $p$ is to prove that we have the right to be sure that it is true. Of course a person may have the right to be sure of something which is false, but we can say 'He has the right to be sure of $p$, but is mistaken' only if we have evidence showing that $p$ is false, evidence which would, if he had it, affect his right to be sure. That is why I cannot say 'I have the right to be sure of $p$, but I am mistaken'. So if we, people generally, have the right to be sure of $p$ it follows, logically, that we have good reason for believing $p$ to be true, and no outweighing reason for believing $p$ false. To show that we have good reason for believing $p$ true and no outweighing reason for believing $p$ false is, surely, an acceptable proof that $p$ is true. So if we can show that we, people generally, have the right to be sure that external objects exist we can give an acceptable, although not a logically conclusive (there might be convincing evidence against, of which we are not aware), proof that external objects exist.

The final question is, then, whether we can prove that we have the right to be sure that external objects exist. Clearly we can show that the accepted conditions for having the right to be sure are satisfied in this case for, as we have seen, what the accepted conditions are is determined by when we are and when we are not ordinarily prepared to say, given that $p$ is true, that someone knows $p$. The fact that we are ordinarily prepared to say that we know that external objects exist shows that the accepted conditions are, in this case, satisfied. The sceptic's argument must be that we cannot show that the information we accept as giving us the right to be sure, and therefore as providing a proof, that external objects exist is in fact correct. That is, we have certain evidence or grounds for saying that external objects exist, and the fact that we all ordinarily say, on this evidence, that we know that external object exist shows that this evidence satisfies our ordinary conditions for having the right to be sure. But in order to show that we do have the right to be sure we still have to show that we do have this evidence, i.e. that this evidence is correct. Now it is a familiar point that any conclusion can be challenged by questioning the premisses from which it is derived. So if I try to demonstrate that certain factors show that we know something (that these factors both give us the right to be sure and, consequently, provide an acceptable proof of the truth of the thing in question) it is always possible for someone to ask 'But how do we know that these factors hold?'

This is the last refuge of the sceptic about our knowledge of the external world. His argument is that we have not proved that we know that external objects exist until we have shown that those factors which we accept as providing us with the knowledge do in fact hold good, until we have shown that what we accept as evidence for the existence of external objects, and as giving us the right to be sure that external objects exist, is in fact the case. We have the argument: $p$; $p$ gives us the right to be sure that external objects exist; therefore we have the right to be sure that external

PERCEPTION AND KNOWLEDGE 157

objects exist. Given our ordinary concept of knowledge the second premiss cannot be challenged, but the sceptic is asking us to prove the first premiss, to prove that $p$ is true. It is obvious that this argument can be extended. If we manage to establish the truth of $p$ the sceptic will then ask us to establish the truth of whatever it was, $q$, that enabled us to establish $p$. And so on. This scepticism— which for want of a better name I will call 'regressive scepticism'— can only be countered if we can come, in the end, to something for which no further proof is needed, something which cannot be doubted, something which will provide an absolutely certain foundation starting from which it can be proved that we do, after all, know that external objects exist.

This is one of the motives that can lead us into a search for the 'foundations of knowledge', but before we embark on the search there is another motive to be considered.

## 10.6 EMPIRICISM

We now turn to the question of *how* we know that external objects exist. How does my perception give me the right to be sure that this, which I perceive, is an external object, continues to exist even when I am not perceiving it? To consider this is to consider, in part, the theory of Empiricism, the theory that all our knowledge and all our concepts come, in the last analysis, from perception, from what we perceive. So let us begin by asking what the theory is, what precisely is meant by saying that perception is the source or origin of all our knowledge, in particular our knowledge of the real existence of what we perceive.

We might mean that this knowledge is such that we would not have it if we did not perceive things. This seems indisputable. Innate knowledge, in the sense of knowledge which we would have whether we ever perceived anything or not may be logically possible, but I doubt whether anyone holds that we do have such knowledge. No doubt it is because we perceive things in the first place that we acquire any knowledge at all, but there is more to Empiricism that that. Rather Empiricism tries to relate specific pieces of knowledge[1] to specific cases of perceiving things. Thus it might be argued that our knowledge of the external world comes

---

[1] By a 'piece of knowledge' I mean knowledge of some particular fact. The questions of what facts are, of what counts as one or two, and so on, are not relevant here.

from our perception in that for any piece of such knowledge there are certain cases of perceiving certain things which are necessary if we are to acquire that knowledge, i.e. such that we could not have the right to be sure of the fact in question unless we had perceived the objects in question. This seems obviously false. It may be that I have to perceive something if I am to have the right to be sure of $p$, but there isn't any particular thing which I have to perceive. I can acquire the right to be sure that, e.g. Cleopatra died from a snake bite in many different ways—by reading about it, by being told about it, by seeing it in a film, by being present myself—and no one of these is necessary.

However this last thesis can be modified to ring very plausibly. Instead of saying that perception is necessary in the sense that we *couldn't* have the right to be sure unless we perceived certain things we might say that perception is necessary in the sense that we *wouldn't* have the right to be sure unless we had perceived certain things. The right to be sure could have been acquired in some other way but, as it happened, we acquired it in this way, and the perception of these things was necessary for our acquiring the right to be sure in this particular way. Knowledge where perception is necessary in this sense, where I wouldn't know what I do unless I had perceived certain things which I have, I will call *sense knowledge*, or knowledge *based on* perception. The suggestion is, then, that our knowledge of the external world, in particular my knowledge that what I perceive is an external object, is senseknowledge, knowledge based on perception. This seems plausible, but it is not, as it may appear, a tautology. It is, in a way, a tautology, and a trivial one at that, that my knowledge that I am perceiving a flower is sense knowledge, in that I wouldn't know that I am perceiving a flower unless I were perceiving a flower, because if I were not perceiving it the statement would not be true. But my knowledge that the flower really exists does not depend on my perception of it in this trivial way, and it is with my knowledge that it really exists that we are here concerned.

This claim that all knowledge is sense knowledge is a main part of Empiricism, but to stop at that would be to stop with a rather weak form of the theory. For the Empiricist maintains not just that perception is a source of knowledge, but that perception is the sole source of knowledge. I shall state this stronger, more debatable, claim as that all knowledge is *derived from* perception. This means that perception is not just necessary, in the sense explained, for the acquisition of the knowledge, but that it is sufficient for the right to be sure, i.e. that it *follows* from the fact that someone has perceived certain things, not necessarily the same things in every case, that he has the right to be sure of the fact in question.

Now we are used to philosophers telling us that this, that or the next thing is really, ultimately, or in the last analysis, of some particular kind, when what they mean is that *they* are not prepared to accept anything as this, that or the next thing unless it is of this particular kind. Is the Empiricist, perhaps, a sort of sceptic who refuses to accept anything as knowledge unless it is based on and derived from perception? Perhaps we can deal with the Empiricist as we dealt with the sceptic, saying that he can refuse to call it knowledge if he wants to, but this is what we ordinarily mean by 'knowledge', and although he may be right in his own special use of the word we can see no reason for changing the meaning just to suit him. I am inclined to think that the boot is on the other foot, for it looks very much as though we will not be able to produce a counter-instance to the Empiricist's thesis, i.e. something which we all ordinarily accept as knowledge which is not at least based on perception, for the very reason that in the ordinary sense of 'knowledge' we do not allow that there can be knowledge of the external world which is not sense knowledge. This is the truth behind the frequent appeal 'But how else could I know, acquire the right to be sure, except from perception?'. It seems that as things are we wouldn't allow that someone did have the right to be sure unless we could cite something the perception of which gave him his right to be sure. This means that Empiricism is a descriptive epistemology, its truth guaranteed by the ways in which we are ordinarily prepared to use the word 'know'. The qualification 'as things are' is important; other ways of knowing may be logically possible but, as things are, perception is the only way of knowing about the external world that we allow that we have.

So this 'How else could we know?' argument seems to give the weaker Empiricist thesis, that knowledge of the external world is *based* on perception, its considerable plausibility. But the fact, granted that it is a fact, that we don't allow that a person knows something about an external object unless he has perceived something (not necessarily that particular thing) which can be accepted as providing him with this knowledge, does not show that perception is the sole source of knowledge. It shows only that perception is necessary for the right to be sure; it does not show that it is sufficient. Indeed it is difficult to see how, precisely, we could prove that perception is the sole source of knowledge. We might show how all our knowledge of the external world *could* come from perception alone. This would not show that it *does* come from perception alone. But if no other suggestions are forthcoming, showing that, and how, all our knowledge could come from perception alone would be as good a proof that it does as we could reasonably hope for.

So the question is whether we can show that knowledge of the external world is derived, or at any rate derivable, from perception. And our own particular topic is: does perception by itself give us the right to be sure that the objects we perceive exist independently of us, exist even when we are not perceiving them? Now clearly just perceiving an object is not, by itself, sufficient to give us knowledge that it exists, or even knowledge of what it is like. To know that it exists, or to know what it is like, I would have to know that my perception is veridical, and my present perception, by itself, does not give me the right to be sure of that. I am not saying that I do not have the right to be sure that this is a table, and that it exists independently of me. All I am saying is that it is the whole history of my past perception, and the context of my present perception, which gives me the right to be sure of it, rather than my present perception by itself.

It seems that if we are to show that our knowledge of the external world, in particular our knowledge that the things we perceive are external objects, is derivable from our perception, we will have to begin by discovering some form of knowledge which is derivable solely from our perception of the particular moment,[1] and then show how our other knowledge can be derived from it. The suggested proof of Empiricism will take the form of starting off with some specially basic form of sense-knowledge, and then showing how our knowledge of the external world can be built up from this basis.

[1] What counts as perception 'of the particular moment'? Any answer to this question is bound to be arbitrary, and I don't think it matters much how we answer it. Perhaps the most convenient, though also the most artificial, answer is to think of the continuous process of perception as consisting of an infinite number of non-extended moments, in the way that a geometer thinks of a continuous line as consisting of an infinite number of non-extended points.

## 10.7 THE FOUNDATIONS OF EMPIRICAL KNOWLEDGE

Thus the attempt to show that we know that objects exist independently of us, and the attempt to show how we know it, both seem to point in the same direction. If we can discover some basic form of sense-knowledge which is both absolutely certain and derived solely from our perception on the particular occasion, then we can provide a final answer to the 'regressive scepticism' discussed in 10.5, and also begin the proof of Empiricism mentioned in 10.6. The search for this basic form of sense-knowledge is usually known as the search for the foundations of empirical knowledge. It is not a search which everyone thinks necessary or advisable.

'The pursuit of the incorrigible is one of the most venerable bug-bears in the history of philosophy'.[1] This pursuit was, in large part, motivated by the mistake of thinking that only what must be true can be known, or of thinking that we cannot have the right to be sure of something unless that something is absolutely certain. But it may also be motivated by the desire to find an answer to a regressive scepticism, not through the mistaken belief that only what is absolutely certain can be known, but because only if, in the end, we can arrive at something which cannot be questioned will we be able to prove an unanswerable proof of how, or even that, we know what we do. Nevertheless not everyone is worried by the regressive sceptic. Some simply accept it as a fact that no absolutely certain foundation is possible, just as it is a fact that absolute certainty is seldom, if ever, possible in the realm of the empirical: 'One seems forced to choose between the picture of an elephant which rests on a tortoise (What supports the tortoise?) and the picture of a great Hegelian serpent of knowledge with its tail in its mouth (Where does it begin?). Neither will do. For empirical knowledge, like its sophisticated extension, science, is rational, not because it has a *foundation* but because it is a self-correcting enterprise which can put *any* claim in jeopardy, thought not *all* at ot once'.[2]

Now it may be that there is no answer to the regressive sceptic (in fact I think there is, but this is something we will come back to). But this does not turn the search for the foundations of empirical knowledge into a hunting of the snark. There is still something to

---

[1] Austin II, p. 104.          [2] Sellars I, p. 170.

be sought, be it snark or boojum. I see a pig; I know that its a pig, and I know that it exists independently of me. But the philosophically important question is *how* do I know it, what gives me the right to be sure. Austin[1] has challenged the idea that there is something that tells me that what I see is a pig. His argument depends mainly on a point about evidence. He insists that seeing the pig is not evidence that its a pig; indeed if I see it, then I don't need any evidence. Now it may well be that this involves an unorthodox use of the word 'evidence', though I think even that could be disputed. But absolutely nothing turns on the choice of that particular term. The question is whether, when I see the pig, there is anything that tells me that it is a pig. Or to put it in a less misleading way: whether there is any answer to the question 'How do I know that its a pig?'. I don't see that anyone can deny that this question has an answer, even if we don't want to describe the information contained in the answer as 'evidence'.

The question I am interested in is not the question of how I know that its a pig, but the question of how I know that its an external object, something which does not depend on my perception for its existence. I believe this question has an answer too, and certainly an answer is worth looking for. This is the philosophical problem about our knowledge of the external world, and to say that we neither have nor need evidence for the independent existence of objects is not to solve the problem. It is to refuse to face up to it.

Thus the quest for the foundations of empirical knowledge is the quest for a final answer to the question 'How does perception provide us with knowledge, give us the right to be sure, of this or that fact?'. These 'foundations' will be pieces of knowledge where the perception of the particular moment is by itself sufficient to give us the right to be sure. It is not necessary that these foundations also be pieces of knowledge about which we cannot be mistaken, but if it turns out that they are then we will have not only an answer to the question 'How do we know?', but also an answer to the regressive sceptic. It is commonly held that these foundations consist in sense datum knowledge, knowledge of the sense data which we immediately perceive. The claim is, first, that sense datum knowledge is knowledge where the perception of the particular moment is, by itself, sufficient to give us the right to be

[1] II, pp. 115-16.

sure, and secondly, that sense datum knowledge is knowledge where we cannot possibly be mistaken. To see whether either of these claims are correct we have first to discover what sense data are, and are like.

## CHAPTER 11
## SENSE DATA

### 11.1 THE NEED FOR A SENSE DATUM TERMINOLOGY

The notion of a sense datum has always been important in the philosophy of perception, but it is not always clear why. There seem to be three possible reasons for the introduction of a sense datum terminology. Notice, however, that none of these reasons show that a sense datum terminology is *necessary*, and although critics sometimes suggest it I do not know of anyone who has wanted to say that a sense datum terminology is necessary. Indeed it is doubtful whether any terminology, as such, can be necessary. It may be that certain terms are necessary *if* we are to say or do certain things, but we can still avoid these terms so long as we decline to say or do these things. Moreover a sense datum terminology would only be necessary, even in the more limited and helpful sense of being indispensable for the making of certain philosophical points or the stating and solving of various philosophical problems, if there were no other way of making those points or stating and solving those problems. And if that were so there would be no way of explaining or introducing the term 'sense datum' in the first place unless, as seems unlikely, there is some adequate ostensive method of explaining and defining it. For if the sense datum terminology can be explained only by translating it into terms we already use and understand, then the points and problems to be raised in the sense datum terminology can also, if less conveniently, be raised, via the same translations, in terms we already use and understand. So the claim must be not that the sense datum terminology is necessary or philosophically unavoidable, but that it is more convenient, less cumbersome, than any ordinary language equivalent. It may enable us, for example, to avoid the vagueness or, even worse, the continual amplification and qualification of a word like 'appears'.

The first reason for introducing a sense datum terminology is that which seems to have motivated Moore. One way of trying to decide between Realist and non-Realist theories of perception would be to sort out what, precisely, a percept would be, and then to consider whether, in fact, we always do perceive such things.

That is, we might isolate that in what we perceive which would be the percept if we were perceiving one, and then ask whether in fact it is a percept. So we distinguish, in what we perceive, the sense datum and ask, as Moore did, 'Does it continue to exist when we stop perceiving?', 'Is it part of the surface of a material object?', etc., i.e. 'Is it, the sense datum, a percept?'. But unfortunately this new question 'Is the sense datum a percept?' is no easier to answer than the original 'Do I perceive a percept?', and we may even be misled into treating 'sense datum' as synonymous with 'percept'—which destroys our original reason for talking about sense data in the first place. The fact is that if the only point of introducing a sense datum terminology, as distinct from talking about percepts, were to enable us to discuss percepts without assuming their existence and so, as Moore hoped, provide some more or less empirical method of determining whether or not we do always perceive percepts, then there would be very little point in introducing it. For not only does this move fail to remove the original difficulties, it raises new ones as well—the familiar difficulties of explaining what sense data are. All that the introduction of sense data can, from this point of view, be said to do is provide a theory-neutral way of referring to the facts, and so show that the facts are theory-neutral, i.e. can be described without assuming any of the philosophical theories of perception.

Perhaps the most common way of justifying the introduction of a sense datum terminology is by means of a version of the traditional argument from illusion: things often, indeed usually, appear different from what they really are and we need, for various philosophical purposes, to distinguish the way things appear from how they actually are. So, it is argued, we need the term 'sense datum' as a name for the appearance, as opposed to the real nature, of what we perceive. But yet remember that we discussed these problems in Chapter 6 without making use of the term 'sense datum' at all, and without raising the perennial problems of what sense data are and what they are like. It seems obvious that the simplest course is to use our ordinary expressions 'looks', 'appears', 'seems', etc., rather than introduce a new and difficult terminology. These ordinary terms may stand in need of some clarification and explanation, but quite apart from the point that it seems we will have to use them to explain the sense datum terminology itself, it does seem that the ordinary terminology will be simpler and less obscure than the suggested 'language of sense data'.

However some have felt that the 'language of appearing' cannot do full justice to the facts of hallucination.[1] Their argument is that since the answer to the question 'What appeared to Macbeth as a dagger?' is 'Nothing' we cannot use the ordinary 'appears' terminology to explain or describe what Macbeth saw. This argument is a consequence of applying the definitional stop to 'perceive', of insisting, as a matter of definition, that we cannot perceive what does not exist. But the fact is that Macbeth did see, in a perfectly straight-forward and ordinary sense, certain expanses of colour, and these expanses looked like a dagger. Of course they did not really exist—that is what we mean by calling it a hallucination—but the 'language of appearing' can still describe what happened, so long as we do not let the stipulative definition that only what exists can be perceived, mislead us into thinking that when a person suffers a hallucination he is not perceptually aware of anything at all.

I think I must conclude, then, that the argument from illusion does nothing to show that the sense datum terminology is helpful in dealing with philosophical questions about the nature of illusion, hallucination and the like. But perhaps this conclusion is unfair to those, like Ayer, who have based the introduction of a sense datum terminology in the argument from illusion. Rather Ayer's argument seems to be something like this: if we are to do justice to the fact that things are not always as they are perceived to be we must make use of a sense of 'perceive' in which it does not follow from the fact that something is perceived that it really exists, or is really as it is perceived to be. In this sense of 'perceive', it will not be true that we always perceive external objects, so let us use 'sense datum' as the generic name for everything that is perceived in this sense of 'perceive'. Despite reservations about talk of different senses of 'perceive' this argument is, I think, untouched by Austin's attack in *Sense and Sensibilia*. In some sense of 'perceive', the ordinary sense as I believe, it does not follow from the fact that a person perceives an $x$, or a thing that is $y$, that there really exists an $x$, or that something is really $y$. But if our knowledge of what really exists is derived from perception the epistemologist must explain how we get from what we perceive to what is

---

[1] Cf. Chisholm II, pp. 116–17.

really there, and in order to do this we need some way of referring to and talking about what we perceive without saying or implying anything about what really exists. This is where the sense datum terminology comes in.

So it seems to me that the real value of a sense datum terminology lies not in connection with the traditional theories of perception, but in the fact that it enables us to raise traditional epistemological and conceptual problems without referring to those theories of perception at all. The introduction of a term which will do the job done by 'idea' or 'sensation' in the past, without its following by definition that there is some entity which exists only in so far as it is perceived, allows us to restate points made by philosophers in the past without committing ourselves to their non-Realist theories of perception. Instead of asking whether all knowledge is knowledge of ideas we can ask whether all knowledge is knowledge of sense data, or more plausibly whether all knowledge can be derived from knowledge of sense data, and instead of asking whether all our concepts are derived from ideas we can ask whether all our concepts are derived from the sense data we perceive.

Moreover there is the suggestion that sense datum knowledge is absolutely certain knowledge, knowledge such that it is not even possible that we are mistaken. If this is so, and if our knowledge of the external world can be derived from sense datum knowledge, we will have found an answer to the regressive sceptic of 10.5. There is even the hope that this certainty of sense datum knowledge might be transferred to knowledge derived from sense data, thus providing us with a final answer to Cartesian scepticism. This has been one of the main hopes of Phenomenalism, although I feel it makes the mistake of trying to meet the sceptic on his own ground, on his own standards of knowledge. Once it is understood that an ordinary empirical statement like 'There is a table in front of me' is not absolutely certain in the sense explained, there seems something rather odd about the attempt to show that it is absolutely certain after all.

However the main point is that it is in epistemology, in the attempt to explain how our perception provides us with knowledge, that the notion of a sense datum is important. Just how, and to what extent, it is important will be more obvious once we have explained what a sense datum is.

## 11.2 MOORE'S METHODS OF EXPLAINING SENSE DATA

Moore's six ways of introducing and explaining the term 'sense datum'[1] seem to include all those commonly used by other philosophers, except Ayer's method of defining sense data in whatever ways seems convenient (we discussed this in 3.3). Most of Moore's methods implicitly rely on some prior understanding of what is meant by 'sense datum'; they are not sufficient in themselves to explain the term. Take, first, the 'method of selection'. The famous instructions in the *Defence of Common Sense* run:

'In order to point out to the reader what sorts of things I mean by sense data, I need only ask him to look at his own right hand. If he does this, he will be able to pick out something (and unless he is seeing double, *only* one thing) with regard to which he will see that it is, at first sight, a natural view to take that that thing is identical, not, indeed, with his whole right hand, but with that part of its surface which he is actually seeing, but will also (on a little reflection) be able to see that it is doubtful whether it can be identical with the part of the surface of his hand in question. Things *of the sort* (in a certain respect) of which this thing is . . . are what I mean by sense data'.[2]

But unless one perhaps accepts some form of a percept theory one is unable to identify anything such that it is doubtful whether it can be identical with the part of the surface of the hand that one sees. If there is something to be 'picked out' other than the surface of the hand we still need to be told what it is, as well as the precise respect in which sense data are things of the same sort as this something.[3]

The 'method of after-images' consists in trying to give an example of a sense datum by reference to cases of our seeing after-images (presumably we might also do so by reference to hallucinations or double-vision). But as several writers have pointed out the difficulty is to discover something in ordinary veridical perception which is like the after-image, e.g. moves as we move our eyes in the way that after-images do. In fact since the point of fastening on after-images as examples of sense data seems to be

---

[1] These six methods are distinguished and named in White I, ch. 8.
[2] p. 54.
[3] For the classic attack on this account of sense data cf. Bouwsma I.

that they do not exist unperceived the natural reply to the question 'What, in what I now perceive, corresponds to the after-image?' seems to be 'Nothing', on the grounds that the things I at present perceive, unlike the after-image, do really exist.

In the 'method of the ultimate subject' Moore explains that the sense datum is that which is the ultimate, real, principle, subject of a sentence like 'This (which I see) is an inkbottle'. Moore agrees that the natural thing to say is that the inkbottle is the subject of the sentence, but he insists that the 'This' also refers to something else, a sense datum. But this cannot tell us what a sense datum is. We have to know what a sense datum is before we can distinguish the reference to the inkbottle from the reference to the sense datum, or even appreciate that the 'This' is doubly referential.

The 'method of intentionality' is slightly more helpful. It was a cardinal point in the *Refutation of Idealism* that all perception must have an object, must be perception of something, and to this Moore adds that all objects of perception, veridical or non-veridical, must have something in common, this common element consisting in some relation with a sense datum. This brings out the point that sense data are involved in all perception, but until we discover what this common element is we still do not know what sense data are.

The 'linguistic method' attempts to explain what is meant by a sense datum by referring to some commonly understood linguistic idiom, usually 'appears', 'looks' or 'seems'. In recent years discussion of sense datum theory has usually taken the form of an examination of these various ordinary language idioms, and some philosophical variants on them, in an attempt to find which one, if any, provides a translation for a sense datum terminology. Personally I doubt whether there is any ordinary language expression which can do the precise job expected of a sense datum terminology—if there were there would be no need to introduce a special terminology. The idioms we ordinarily use, and philosophers' variations on them, normally carry one of three undesirable implications: that what is perceived does really exist, that it does not really exist, or that there is some special entity which comes between us and what we would ordinarily describe ourselves as perceiving. By far the best course seems to be to use the ordinary 'is' idiom—plus any of the other we might happen to need—but to

F*

make it quite clear what precisely is being said and implied when those idioms are used in sense datum statements.

However let us take a brief look at some of the relevant expressions, remembering that they can be used in a variety of contexts and with a variety of implications, such that their precise force varies considerably from occasion to occasion. All we can do here is indicate the most natural or most common ways in which a particular expression might be taken. The simple 'It is a dagger' ordinarily implies, although I have argued that it does not entail, that there really is a dagger there, but once this implication is explicitly denied the 'is' terminology is the best for our purposes. On the other hand 'It is "a dagger"' (cf. I 'see' a dagger; I see 'a dagger') naturally implies that there isn't a real dagger there. 'It is an apparent dagger', 'It is a seeming dagger' and 'It is an appearance of a dagger' all suggest that I am describing some special sensory object, not to be identified with the real dagger should one happen to be there. 'It seems to be a dagger', 'It looks like a dagger' and 'It appears to be a dagger' all suggest either that it is not a real one or that I am not quite sure whether it is a dagger or something else.

These last idioms are the most favoured when people try to explain sense data, and we are often told that the sense datum is the way that what is perceived appears to the perceiver. I do not think this is at all helpful. First of all there is the difficulty of determining the precise sense of 'appears' involved and we shall see that the sense datum is not to be identified either with the appearance of what we perceive or with what we take ourselves to perceive. There is also the point that talk about the 'appearance' or 'look' of things may lead us to treat the sense datum as some thing distinct from that of which it is the appearance or look. Nor is it clear from this account what would count as a description of a sense datum, a sense datum statement. The tilted penny looks elliptical but it also looks circular—does that mean that the sense datum, the look of the penny, is both round and elliptical? And if, as usually happens, I fail to notice the look of the penny, that it looks elliptical, does this mean we can, after all, perceive without perceiving a sense datum? Or are we to say that the sense datum is what what is perceived looks like? If so, am I to say, when I see an inkblot that looks like a dagger, that the sense datum is a dagger? And the inkblot also looks like an inkblot! And what if someone says to me 'I don't want to know what the sense datum is like; I want to know what it *is*'? In describing our sense data we sometimes want to *identify* what we perceive and the 'looks' terminology makes it pretty well impossible to do this. On the other hand we may want to say what the sense datum is like, as well as saying what features it actually has, and then we will want to use the 'looks' terminology as well as the 'is' terminology. We will have more to say on this subject in 11.7.

Recently White[1] has suggested a variant on the linguistic method, taking over Hampshire's[2] talk of 'non-committal descriptions'. The suggestion is that a sense datum statement is one which 'describes' what is perceived without 'identifying' it (cf. 6.2). But Hampshire's 'non-committal' descriptions are non-committal in precisely the wrong way. Hampshire is interested in cases where the perceiver is not sure what it is that he perceives, whereas what sense datum statements should be non-committal about is the real existence of what is perceived.[3] Often we do want to say of what figures in a sense datum that it is a thing of a certain kind—an expanse of scarlet, a shriek, a pungent smell. What makes White's suggestion plausible is that identification usually involves naming or referring to a physical object and, it is normally argued, sense datum statements cannot name or refer to physical objects. But even so we must distinguish *identifying* what we perceive from saying whether it *really exists*, and it is the latter, not the former, that is excluded as such from sense datum statements.

Finally there is the method I shall follow, the 'method of restriction'. This works by restricting what is perceived to a certain limited range of perceptual objects, which are then said to be 'immediately perceived', sense data being what we immediately perceive when we perceive something. It is important to realize that this restriction in the sense of the verb 'perceive' does not mean that we are referring to some special and separate mode of acquaintance, which has to be distinguished from ordinary perception. Philosophers have often spoken as if there were two dis-

---

[1] I, ch. 8.   [2] Cf. I, II.

[3] Hampshire himself does not seem clear about the difference between these two. For example he says (II, p. 84) 'In the Macbeth situation we are to suppose that the speaker is altogether unable to identify the phenomenon to which he is referring as a thing of a specific kind' when it seems obvious that Macbeth knew what sort of thing it would be, but was unsure whether it really existed.

tinct processes, awareness of sense data and perception of physical objects, but this duplication is quite unnecessary and grossly implausible. We would much rather say that we immediately perceive a red sense datum in perceiving a tomato and, for that matter, perceive a tomato in immediately perceiving the red sense datum. Immediate perception is not a special and separate form of perception, it is ordinary perception of a specially restricted range of things, just as lieder singing is not some special and separate form of singing but ordinary singing of a specially restricted range of songs. For this reason it might be preferable not to speak of a special sense of 'perceive' at all, but on the other hand some things which are true of immediate perception could not be true of non-immediate perception, and vice versa. The point is not that 'perceive' has different senses, much less that there are different types or processes of perception, but that 'immediately perceive' has a different sense from 'perceive' *tout court*. The difference between them lies in what can correctly be said to be perceived or immediately perceived.

## 11.3 IMMEDIATE PERCEPTION

To explain sense data, then, we have to explain what can and cannot be immediately perceived. Immediate perception is not to be identified with sensory awareness—sense data are not to be defined as what we sense. For when a person perceives a tree he also senses a tree, whereas it is usually insisted that physical objects cannot, as such, be sense data. Moreover it would be silly to speak of the sense datum as what is sensed as opposed to what we have perception-that of, for precisely the same things are involved in each case, i.e. perception-that involves making 'judgements' about what is sensed. And, finally and conclusively, immediate perception must involve some element of perception-that if we are ever to say or know anything about our sense data. There may be some tendency to say that strictly speaking we sense nothing but sense data, just as there is a tendency to say that strictly speaking we perceive nothing but sense data, but if we are to describe our sense data there will have to be something more to our perception of them than mere sensory awareness. And any attempt to explain the sense in which we sense nothing but sense data will merely reproduce our discussion of immediate perception, of the sense in which we perceive nothing but sense data.

## SENSE DATA

The crucial point about immediate perception is that it does not go beyond what is perceived at the particular moment. Or, as we might prefer to put it, sense datum statements, statements describing what we immediately perceive, do not refer to or describe or entail or imply anything about anything which is not being perceived, in its entirety, at that particular moment. This point has sometimes been stated in a rather misleading way. Warnock, for example, talks[1] about statements which involve no 'inferences or assumptions'. Now suppose I were to say that I see a bed in which Queen Victoria once slept. This would not describe what I immediately perceive because although I am talking about what I now perceive I am also referring to, and making an assertion about, something which I do not now perceive, viz. Queen Victoria. Even to describe what I perceive as a bed is to go beyond what I perceive at this particular moment, for to say that it is a bed is to imply, perhaps entail, that it has a bottom and a back, neither of which I now perceive. But this does not mean that I am inferring or assuming anything when I say that this is a bed in which Queen Victoria once slept. Despite what some philosophers suggest[2] I certainly do not infer that this is a bed, i.e. look at the top and side facing me and then conclude, consciously or unconsciously, that there will be another side, a back and a bottom behind what I now see. The simple truth is that I see it *as* a bed and, in all probability, do not pause to distinguish what I do see of it from what I do not.[3] Similarly to say that I assume that this is a bed in which Queen Victoria once slept would be misleading, because it suggests either that I am mistaken in thinking that it is, or at least that I do not have the right to be sure, and hence do not know, that it is. But I do not assume that this is a bed in which Queen Victoria once slept, I *know* that it is. Nevertheless, I would insist, this knowledge goes beyond what I now perceive, and this is the truth behind the misleading talk of 'inferences and assumptions'.

In what ways do statements of immediate perception, sense datum

---

[1] I, chs. 8 and 9. He also talks about statements which 'take nothing for granted'; this is preferable but not, perhaps, particularly clear.

[2] Cf. Ayer on 'errors of inference', II, pp. 38 ff.

[3] Walton I argues that anything which can be said to be inferred from what is perceived can also be said to be perceived as such, without inference. If it is possible for one person to infer from what he perceives that it is *x*, then it is possible for another person to perceive it as *x*—it is all a matter of training, experience and 'perceptual dispositions'.

statements not go beyond the perception of the particular moment? We have already mentioned the first point, that statements of what is immediately perceived say or imply nothing about the real existence of what is perceived. Nor do they say or imply anything about the real nature of what is perceived. We have seen that this is not equivalent to saying that sense datum statements cannot identify what is perceived, in the sense of saying what sort of thing it is that is perceived, although there will be limits to what identifications can be made in a purely sense datum statement. Nor is this equivalent to saying that sense datum statements are about what must exist if it is perceived, although in so far as it is true that sense datum statements are about what we perceive it will be true that sense datum statements are about what must have perceived existence. For Price,[1] and Ayer,[2] a sense datum is something which must exist if it seems to exist: 'When I see a tomato there is much that I can doubt. I can doubt whether it is a tomato that I am seeing and not a cleverly painted piece of wax. I can doubt whether there is any material thing at all. Perhaps what I took for a tomato was really a reflection; perhaps I am even the victim of some hallucination. One thing however I cannot doubt; that there exists a red patch of a round and somewhat bulgy shape, standing out from a background of other colour patches, and having a certain visual depth, and that this whole field of colour is directly present to my consciousness . . . Data of this special sort are called *sense-data*'.[3] All this is quite mistaken. If I am suffering from a hallucination then there is no more a red patch of a round and somewhat bulgy shape than there is a tomato. And if I can doubt whether it is a tomato and not a cleverly painted piece of wax, I can doubt whether it is a somewhat bulgy shape and not a cleverly painted piece of flat canvas. But if, on the other hand, all that is meant by saying that I cannot doubt the existence of the red patch is that I cannot doubt that it has perceived existence (because, *ex hypothesi*, I perceive it), precisely the same is true of the tomato. In so far as sense data are perceived they must have perceived existence, but this is true of anything. The special point about sense datum statements is that they say or imply nothing about *real* existence.

Next, immediate perception is such that our sense data include only what is perceived in its entirety at the time in question by the

[1] I.      [2] III.      [3] Price I, p. 3.

sense in question; sense datum statements say or imply nothing about what is not, at that moment, being perceived. If I see a bump in a tablecloth caused by a jug hidden underneath it I might be said to see the jug, in that I see the bump and know that the jug makes it, but I cannot be said to immediately perceive the jug. Nor is anything that is perceived 'indirectly', in the Causal theorist's sense, immediately perceived. To hear a train is to hear a sound made by the train, and in the direct-indirect terminology my hearing the train is indirect perception in that I hear it only in the sense that I hear an effect of, something produced by but distinct from, the train itself. The sound, on the other hand, is directly perceived. This means that the train is not immediately perceived either, at least not via the sense of hearing. The sound is what is immediately perceived. Nevertheless immediate perception is not to be equated with direct perception. Even if I bring the jug out from under the cloth, and see it 'directly', I do not immediately see the jug, because I do not see all, every part, of it. The most that I see is the outside surface facing me; I do not see its other sides, or its inside, let alone the interior of its sides. It would be stupid to conclude from this that I do not see the jug, for to perceive every part of the jug I would have to break it up into pieces so thin as to be transparent! What I now see of it is just what counts as seeing the jug.[1] But we can and must conclude from this that I do not *immediately* see the jug, for this is part of what is meant by 'immediate perception', i.e. I immediately perceive $x$ only if, at the time in question, I perceive all, every bit, of $x$. If I do not see all of $x$ then to say that I see $x$, no matter how correct and no matter how justified, is to go beyond what I now see, to talk about bits and parts that I do not now perceive, and so to go beyond a report of immediate perception, a sense datum statement.

There is also a third point. Not only does hearing a train amount to hearing a sound produced by, an effect of, the train, but also we hear trains, or other objects, only in so far as we hear sounds belonging to them. Similarly with the other senses. We taste things, for example, only in so far as we taste tastes, and we smell things other than smells, e.g. roses, only in so far as we smell smells belonging to them. There is for each sense modality a type of object (the 'sense object' of that sense, as I will call it)

[1] Cf. Warnock II.

which it is the function of that sense to acquaint us with.[1] This is even true of sight and touch, although they differ from the other senses in that their 'sense objects' are not items distinct from, and belonging to, physical objects, in the way that sounds or smells are distinct from, but belong to, physical objects. The sense objects of sight are 'coloured expanses'; it is impossible to see something unless in seeing it we see a coloured expanse (the sense objects are, in effect, the 'tautologous objects' of the various senses, cf. 5.1). These coloured expanses may be two-dimensional patches of colour, they may be three-dimensional volumes of colour, but most commonly they will be the surfaces of three-dimensional objects. Finally there is the sense of touch, which can be divided into the temperature sense and the sense of touch proper. The sense objects of the temperature sense will be degrees of heat and cold, which we can refer to as 'temperatures'. There is no obvious name for the tautologous objects of the sense of touch proper, so let us speak of 'contacts'. A contact is what we feel whenever we feel anything, and includes such features as the texture, the hardness and the pressure of what we feel. So we have this list of the sense objects of the various senses: coloured expanses (sight), sounds (hearing), tastes (taste), smells (smell), heat and cold (the

---

[1] I think this is the point behind the Berkelean remark that, strictly speaking, we hear nothing but sounds, smell nothing but smells, etc. In his discussion (I, ch. 7) Warnock takes the point to be that 'I hear a sound' holds up fewer hostages to fortune, is less likely to be mistaken, than 'I hear a coach'. What Berkeley is looking for, Warnock suggests (cf. pp. 164–74), is an account of what we perceive which cannot be mistaken, and, he argues, we find this in 'It seems to me as if I were hearing a sound'. But this is incorrigible not because it is restricted to what I hear, and makes no 'inferences or assumptions' about what is not heard, but because 'strictly speaking' it is not about what I hear at all. Like 'It seems to me as if I were hearing a coach', or even 'It seems to me as if I were hearing Dean Swift's coach driven by Dr Johnson' (which are incorrigible in precisely the same way, although neither would count as a description of a Berkelean idea), it is about me, or rather about what I take myself to hear. The 'It seems to me . . .' indicates that these statements report my judgement about what I hear, and although what I hear may be quite different from what I think it is, I cannot be mistaken as to what I think it is. Indeed, pace Warnock p. 174 ('This formula serves to make it clear that I *am* having some "perceptual experience"') the truth of these statements is quite compatible with my not hearing anything at all. It seems to the lunatic as if he were hearing God's instructions to him, but in fact he hears, not even as a hallucination, no such thing at all. The point of saying 'Strictly speaking we hear nothing but sounds' is not that we might be mistaken in saying we hear coaches (for we might still be mistaken in saying we hear sounds) but that we hear coaches and such things things only in so far as we hear sounds of them.

SENSE DATA 177

temperature sense), and 'contacts' (touch proper). And notice once again that to say that strictly speaking we see nothing but coloured expanses, hear nothing but sounds, etc., is not to deny that we ever see tables or hear trains. It is only to say that we see and hear such things as tables and trains only in so far as we see coloured expanses or hear sounds belonging to them.

So the third important feature of immediate perception is that it is restricted to the sense objects of the appropriate sense. I can, quite correctly, be said to hear a train or see a table, but what I immediately perceive is not the train or the table, but the sound made by the train, the coloured expanses which comprise the visible surfaces of the table. The point of this restriction is that in order to realize that I am perceiving things other than the appropriate sense objects I have to realize certain facts which I do and could not learn from my present perception alone, e.g. that this noise is made by an object of a certain shape and structure, or that other areas and surfaces, which are tangible as well as visible, lie behind the coloured expanses I do see. Although in a perfectly respectable sense I perceive more than the relevant sense objects, to say that I perceive this more is to go beyond what I now perceive.

Finally, the report of what is immediately perceived describes the sense objects I perceive as I perceive them to be, not as they really are. The sound I hear may, in itself, be loud and squealing but if, due to the cotton wool in my ears, it sounds soft and muffled to me then the correct sense datum description of that sense object is 'Soft and muffled', not 'Loud and squealing'. Obviously my perception of the moment by itself can tell me only how what is perceived is perceived as being, and not how it really is.

So to sum up, a sense datum statement, a description of what is immediately perceived, is an account of what is perceived which:

(1) describes and refers to only a certain range of objects characteristically perceived by the sense modality in question, i.e. the sense objects of that sense modality;

(2) describes and refers to only those sense objects which are perceived at the time in question via the sense modality in question, and to only those parts or aspects of those sense objects which are then being perceived;

(3) describes those sense objects as they are perceived by that

perceiver to be, saying or implying nothing about what does or does not really exist, nor about what the items perceived are or are not really like.

It has been doubted[1] whether there could be statements corresponding to this account of a sense datum statement as a statement which describes what is perceived without going beyond the perception of the particular moment. It is hard to know what reasons can be given, *a priori*, for this scepticism, but I think there is one mistake which might tempt us to this conclusion. Hampshire, for one, seems to think that any account of what we learn from the senses at a particular moment alone, cannot make use of terms which are also used to refer to what is not perceived on particular occasion (cf. the argument that Phenomenalism cannot work because it is impossible to describe sense data without using physical object terms). Yet there is no inconsistency in a sense datum statement's including terms which are normally used to refer to things other than sense data, even in its including physical and external object terms, so long as these terms are modified so as explicitly to remove any physical or external object implications, e.g. by talking about 'table-like' sense data or sense data 'as of' a table.

And, of course, terms used in a sense datum statement, to describe and refer to what is perceived on this particular moment alone, can also have application to things which are not perceived at this particular moment. If we could not, in describing what is immediately perceived, use terms which can also be applied to things which are not immediately perceived on this occasion we could not say anything. The terms by which we classify what we immediately perceive must have application to other things too, and this is permissible so long as what we say in those terms, in the sense datum statement, does not go beyond what is immediately perceived on that particular occasion.

## 11.4 THE NATURE OF SENSE DATA

What, then, are sense data? For the Idealist and the Causal theorist, sense data will in fact (but not by definition) be percepts —private, mental, sense-dependent entities, of the same sort as after-images and hallucinations. But we have decided to accept Realism, and for the Realist sense data will not be entities at all.

[1] E.g. Hampshire II, p. 83, Strawson III, p. 97.

Sense data are what we immediately perceive. But this does not mean that sense data form some special class of objects which are perceived in some special way. Sense data are not, on a Realist interpretation, objects at all, and to talk about sense data is to talk, in a special way, about the things, whatever they may be, that we happen to perceive. In particular it is to talk about what we perceive without going beyond what we perceive, to describe what we perceive only as we perceive it to be, and only in so far as we do perceive it.

Of course we can talk in many different ways about what we perceive. We can, for example, talk about the external objects we perceive, and in talking about them we will be talking about the same things that we talk about when we talk about our sense data. But we will be talking about them in a different way. Except for such special cases as Macbeth's dagger, external objects—or at any rate parts and aspects of external objects—make up our sense data. Thus a large part of the surface of the paper on which I am writing is included in my present[1] visual sense datum.

This way of talking can be misleading. When I say that my present visual sense datum includes part of the surface of the paper, or that the surface of the paper figures in my present visual sense datum, I do not mean that the surface of the paper is part of some complex object called my present sense datum, in the way that a brick may be part of a house or the surface of the paper part of a drawing. I mean only that in talking about my present visual sense datum we are talking about the surface of the paper, although not precisely in those terms. Thus to say 'This sense datum includes $x$', or '$x$ figures in this sense datum', is to say 'The description of this sense datum includes a reference to, or a description of, something which can also be described, although not in a sense datum statement, as $x$'.

There is, therefore, no conflict between 'We always perceive sense data' (i.e. 'Whenever we perceive we immediately perceive') and the Realist's 'We perceive external objects'. Given Realism, to perceive an external object is, among other things, to perceive a sense datum, a sense datum which might be said to include parts or aspects of that external object. And notice that to say that we

---

[1] How long do sense data last? The answer is as in the footnote to p. 160—that it is perhaps most convenient to think of sense data as instantaneous, as what we perceive at some non-extended moment of time.

always perceive sense data is not to say that we perceive nothing but sense data, any more than to say that we always jump jumps is not to say that we can jump nothing but jumps. Rather, we perceive things only in so far as we perceive sense data 'of' them, just as we jump streams, fences and so on only in so far as we jump jumps over them.

This brings out that 'We perceive sense data' is an analytic truth because 'sense datum' is, in effect, an internal accusative after 'perceive'. A failure to appreciate this point, together with a confusion between two senses of 'sensation', invalidates Ryle's attack on sense datum theories. Ryle argues that the sense datum theorist is 'talking the same sort of nonsense as he would if he moved from talking of eating biscuits and talking of taking nibbles of biscuits to talking of eating nibbles of biscuits.... He cannot significantly talk of "eating nibbles" for "nibbles" is already a noun of eating, and he cannot talk of "seeing looks" since "look" is already a noun of seeing'.[1] This argument cannot be valid, for in this sense 'jump' is a noun of jumping and 'gift' a noun of giving, and yet we talk quite significantly of jumping jumps and giving gifts. In so far as 'sense datum' is an internal accusative there is nothing illegitimate about 'I perceive a sense datum', nor, for that matter, about 'I eat a nibble'. This would only be nonsense if 'sense datum' referred not to what is perceived but to the perceiving of it, for it is absurd to speak about 'perceiving perceivings' or about 'eating nibblings', just as it is absurd to speak of 'jumping jumpings' or 'giving givings'. But there is no reason for interpreting 'I perceive a sense datum' in this peculiar way unless we talk, as Ryle unfortunately does, about 'perceiving sensations'. For 'sensation' might mean either 'perceptual sensation' (sense datum) or 'sensing', and only in the former sense is it legitimate to talk of perceiving sensations.

Ryle also argues[2] that if we have to perceive a sense datum in order to perceive something, we will then have to perceive another sense datum in order to perceive that first sense datum, and so on *ad infinitum*. This also relies on the failure to realize that 'sense datum' functions as an internal accusative. Perceiving a sense datum in perceiving a tree is not like buying a book in order to read it, and it no more requires the perceiving of a previous sense datum than the jumping of a jump, which is necessary if we are to

[1] I, p. 217.  [2] I, p. 213.

jump a fence, requires the jumping of a previous jump. The internal accusative refers to something which is logically tied to the occurrence of what is named by the verb, and not to some condition which must be satisfied before whatever it is can occur.

Thus a failure to appreciate the fact that 'sense datum' functions as an internal accusative is partially responsible for the common feeling that the notion of a sense datum is somehow illegitimate. Obviously an even more important mistake is the failure to distinguish between sense data and percepts—as if the analytic truth that all perception is perception of sense data could show that Realism is false. This error of thinking of sense data as percepts is particularly easy to make once we start asking about the nature of sense data, for it is actually possible to argue that sense data are—like percepts—mental, private and even sense-dependent!

For a start, sense data are certainly not physical entities, at least not in the sense explained in 1.2. For although we might want to say that the various parts and aspects of external objects which figure in our sense data are located in physical space, sense data themselves are not. Nevertheless it would be misleading to take this to show that sense data are mental entities, for that suggests that they are entities which can be assigned a location ('in the mind'), albeit a non-physical one. Rather the point is that they are not entities at all.

As for privacy, it is seldom explained whether the privacy of sense data, or of percepts for that matter, is supposed to be a necessary or a contingent matter. The usual suggestion is that since two people cannot stand in precisely the same spot at precisely the same time what they perceive must always be slightly different. But this privacy of sense data will be at best a contingent matter, and if we allow that two disembodied consciousnesses might perceive from the same place, or that some trick of surgery might enable one person to perceive through another's eyes, we also allow that it is possible for two people to perceive the same sense datum. The point should be, I think, that the privacy of sense data is a logical matter, a consequence of the definition, or rather the individuation, of sense data. We could individuate sense data by what they are sense data of, but this is neither helpful nor desirable if we want to think of knowledge of sense data as, somehow, coming before and explaining our knowledge of external objects, what they are sense data of. It would be better to individuate sense data by reference to their qualitative features and the times at which they are perceived. By itself this would mean that if different people perceived qualitatively identical things at the same time, as is surely possible, they would perceive the same sense datum. So we need to add that sense data are also individuated by reference to the people who perceive them, and this does make it logically impossible for two people to perceive the same sense datum. It now follows from the fact that $A$ and $B$ are different people that they are perceiving different sense data; the sense datum that $A$ perceives is what $A$ perceives as $A$ perceives it to be and in so far as $A$ perceives it, whereas the sense datum that $B$ perceives is what $B$ perceives as $B$ perceives it to be and in so far as $B$ perceives it. The reason for making the privacy of sense data a logical matter is, I take it, that our main interest in sense data is epistemological, and since the knowledge I derive from the sense data I perceive will be my knowledge and mine alone, there is some point in insisting that the sense data too are mine and mine alone. My (private) sense data are what give me my own (private) knowledge of the objects in question.

Finally, sense-dependence. Those philosophers, like Moore and Broad, who have been at pains to keep 'sense datum' a theory-neutral term, have often insisted that sense data must be able to exist unperceived, presumably on the grounds that to make this impossible would be to make Realism false. But I think that once we distinguish between the sense datum and what may be said to figure in or be included in the sense datum, we can retain Realism by insisting that the latter may, and usually do, continue to exist unperceived, while at the same time saying that the sense datum itself exists only in so far as we perceive it. Indeed since a sense datum statement says or implies nothing about what really exists, the only sense in which we could talk of a sense datum as existing would be in the sense of perceived, i.e. sense-dependent, existence. It would be extremely confusing, indeed illegitimate, to say that the sense datum really exists. But to say that sense data exist only when perceived will not be to say that the items which figure in the sense datum exist only when perceived. Rather it will be to say that in so far as we describe what we perceive in sense datum terms, we are talking about it only to the extent that it is perceived. Passing from the analytic 'Sense data exist only when perceived' to 'sense data are things which go out of existence when we cease perceiving' would be like passing from the analytic 'No minor is more than twenty-one years old' to 'Minors are people who go out of existence when they reach the age of twenty-one'. That is, what is meant by 'Sense data are sense-dependent' will be merely that things can be referred to in sense datum statements only in so far as they are perceived—although those same things can continue to exist unperceived—just as people can be referred to as minors only in so far as they are under twenty-one—although those same people can continue to exist past the age of twenty-one.

We can, then, insist that sense data are private, sense-dependent and non-physical without committing ourselves to a percept theory of perception, so long as we remember that these sense data are not to be thought of as entities in their own right. But there still remains a strong suspicion that there must be a radical distinction between sense data and external objects, in that certain things which are true of the former are not true of the latter, and vice versa. In part this suspicion stems from a misunderstanding of what sense data are and how they are to be defined, in particular from the attempt to explain sense data by reference to how things appear. If we are told that the sense datum is the way what we perceive appears we may conclude that since the penny looks, in a definite sense, elliptical it follows that the sense datum is elliptical, and therefore that the sense datum cannot, in any sense, include the surface of the penny. The truth is that 'It is elliptical' would be an incorrect sense datum statement, for the sense datum statement describes the brown expanse that is perceived, and to say that this brown expanse is elliptical is simply false. What we can say of the sense datum, as of the penny, is that it looks elliptical. We have already seen (11.2) that the definition of sense data in terms of how what is perceived appears is not only misleading, but also deprives us of the use of the 'looks' terminology which we may well, as here, want to use in a sense datum statement.

Again if I look through a piece of blue glass the sense datum I perceive will be blue. Yet what I perceive is not blue, so it seems to follow that the sense datum does not, in any sense, include parts or aspects of the external object I perceive. Once again talk of sense data seems to lead us away from Realism. But even though the colour I see may not be the colour of the external object it does not follow that I am not perceiving that object, nor immediately perceiving parts of it. For it does not follow from the fact that I see part of the surface of an external object that I see the colour of that surface—indeed that is just what does not happen in our example—nor does it follow from the fact that the colour I see is not that of the external object, that I am not perceiving the external object, nor immediately perceiving parts of it.

## 11.5 SENSE DATUM KNOWLEDGE

Sense datum knowledge is knowledge of the truth, or for that matter the falsity, of sense datum statements. Just as not all statements about sense data qualify as sense datum statements, so not all knowledge about sense data qualifies as sense datum knowledge. My knowledge that my sense datum is a sense datum of the house that Jack built is knowledge about a sense datum, but it is not sense datum knowledge, for 'This is a sense datum of the house that Jack built' is not a sense datum statement, although it is a statement about a sense datum.

Now the question was whether sense data, or more accurately sense datum knowledge, would provide that 'foundation of empirical knowledge' for which we were looking. We were looking for knowledge which was (1) derivable solely from the perception of the particular moment, and (2) absolutely certain, such that we could not be mistaken about it. In fact sense datum knowledge possesses neither of these features.

First of all, the special feature of sense datum knowledge is that it is knowledge that does not go beyond what is perceived at the particular moment. This does not mean that it is knowledge that I would have even if I never perceived anything except what I now perceive. It is doubtful whether I could know anything at all if I perceived at only this one moment—particularly if this moment is a non-extended temporal point! Nor does it mean that it is knowledge of what what I perceive would be like if I never perceived anything except on this occasion. It is quite possible that previous perception modifies present perception, so that what I perceive now is affected by what I have perceived on other occasions. What it does mean is that sense datum knowledge is the *most* that the perception of the particular moment *could by itself* give us the right to be sure of. Of course perception can give us the right to be sure of things we do not perceive, as when seeing a dripping umbrella in the hall gives me the right to be sure that it is raining, even though I do not see the rain. But I have the right to be sure of this only because I already have the right to be sure that when there are dripping umbrellas there is also, more often than not, rain. And obviously my present perception does not, by itself, give me the right to be sure of that. Again, if on a hundred occasions I see an egg break when it is dropped I have the right to be sure that if another egg is dropped in similar circumstances it too will break, and so my perception gives me the right to be sure of what I do not perceive. But simply perceiving one egg break when dropped does not, by itself, give me the right to be sure that the next egg will break when dropped. The point is that perception on one occasion cannot, by itself, give us the right to be sure of what is not perceived on that occasion. Perception on one occasion together with other information can give us the right to be sure about what is not perceived on that occasion, and perception on several occasions can give us the right to be sure about what is not perceived on any of those occasions, but knowledge which is derived solely from the perception of the particular moment cannot go beyond what is perceived at that moment.

So because sense datum knowledge is knowledge which does not go beyond what is perceived at a particular time, sense datum knowledge will be the most knowledge that that perception can provide us with. But this is not to say that it is knowledge which is derivable from the perception of the particular moment, knowledge such that the perception of the particular moment is sufficient to give us the right to be sure. I have the right to be sure that I am at present perceiving a red expanse because I know that I never, or very seldom, make mistakes over colours like red, but it isn't my present perception, by itself, that gives me the right to be sure that I seldom make such mistakes. What my present perception, by itself, gives me the right to be sure of is not that I am perceiving a red expanse, but that that is what I take myself to be perceiving.

The point is that if I am to have the right to be sure about what I perceive at this particular moment, I have to have the right to be sure that I do perceive, and don't merely 'perceive', think I perceive, it. And my present perception, by itself, does not give me the right to be sure of that. We might put this by saying that sense datum knowledge is the knowledge which perception of the particular moment does provide us with, given only that we make no mistakes of perception-that, i.e. given that I do not take what I perceive to be other than what it is.[1] And clearly my perception of the particular moment is not by itself sufficient to give me the right to be sure that I make no such mistakes. Sense datum knowledge is knowledge which does not go beyond the perception of the particular moment, but nevertheless it is not knowledge which is derivable from that perception.

One reason for thinking that it is knowledge derivable from the perception of the particular moment might be the idea that we cannot be mistaken about our sense data, that it is impossible, where sense data are concerned, to make mistakes of perception-that. I will come back to this in a moment. Another reason for thinking that perception of sense data must be sufficient, by itself, to give us knowledge of those sense data, might be the old idea of 'knowledge by acquaintance'. The theory is that there is a type of knowledge which consists in being 'acquainted with' what is known. This acquaintance is thought of as some form of perceptual acquaintance; in our terms the suggestion would be that immediate perception constitutes a form of knowledge.

I have heard it argued that the notion of 'knowledge by acquaintance' is misconceived because it makes use of a strange sense of 'acquaintance'—on this theory our acquaintances are not friends and relations, but coloured expanses, sounds and smells. A less trivial objection is that it makes use of a strange sense of 'knowledge': 'There is no common sense of "know" such that from the mere fact that I am seeing a person it follows that I am at that moment knowing him'.[2] It just isn't clear how perceiving, apprehending, being perceptually acquainted with, something could possibly count as knowing it. No doubt one source of the theory of knowledge by acquaintance is the familiar but mistaken idea that knowledge consists in some sort of mental process which involves

---

[1] This may seem to let in too much. For if I take what I perceive to be a typewriter sold to me, as once belonging to Shakespeare, by an antique dealer in South Croydon, and if it is given that I make no mistakes of perception-that, it might seem to follow that my present perception gives me the right to be sure that this is a typewriter sold to me, as once belonging to Shakespeare, by an antique dealer in South Croydon. And that, clearly, isn't sense datum knowledge. However, even if I make no mistakes of perception-that, even if I am right in thinking it is such a typewriter, it doesn't follow that my present perception, by itself, gives me *the right to be sure* that it is such a typewriter. My present perception gives me the right to be sure only that I am perceiving a certain sense datum, which I take to be a sense datum of the typewriter.

[2] Moore IV, p. 77 n.

SENSE DATA 187

grasping or apprehending some subject. Given this we might think of perceiving as a kind of knowing, in so far as perception is some sort of mental process which involves 'grasping' or apprehending some object. But knowledge is not a process at all, and certainly does not consist in grasping anything, except in a highly metaphorical sense. We are often told, these days, that memory is not a source of knowledge, as some have suggested, but rather a type of knowledge. To this can be added the reversed point that perception is not a type of knowledge, as some have suggested, but rather a source of knowledge. Perhaps the ambiguity of a phrase like 'way of knowing' has led some to think that if something provides us with knowledge it must itself be a type of knowledge.

Even writers who do not talk about knowledge by acquaintance are inclined to talk of perception as a type of knowledge. Presumably it is his belief that sensing is a type of knowledge[1] that leads Price to define the sense datum as that in what we perceive about which we cannot be mistaken, and it may be the same mistake[2] that leads Hirst to attribute common sense Realism with the 'immediacy assumption', which seems to include the belief that perception is always correct or veridical. It may be that perception cannot be incorrect, but this does not mean that it constitutes, or even provides us with, absolutely certain knowledge of what is perceived. For, although perception may be veridical or nonveridical, it cannot be correct or incorrect—it simply is as it is. To know what we perceive we must be more than just aware of it; we must come to some opinion about what it is, and that opinion must be correct. To say that perception constitutes knowledge because it is as it is and cannot be mistaken is to be like Wittgenstein's man who puts his hand on top of his head and says 'I know how tall I am'.[3]

So we can see how a misunderstanding of what knowledge is, together with a misconstruction of the point that in acquainting us with something perception cannot be mistaken or incorrect, can lead us to think that perception is a form of knowing, and an absolutely infallible one at that. This in turn leads to the view that Realism must be mistaken, since some of things we perceive do not really exist, or are not really as they are perceived to be.[4] It

[1] Cf. I, p. 274.   [2] Cf. II, pp. 244-5.   [3] I, § 279.
[4] Prichard regretted 'the almost universal tendency to take it for granted, without serious consideration, that perception in its various forms is a particular

also leads to the view that sense data 'must necessarily appear in every particular what they are, and be what they appear'.[1]

We now turn to the question of whether sense datum knowledge is absolutely certain knowledge. That is: can we be mistaken about the nature of our sense data? Can my present sense datum be different from what I think it is? The usual answer is that the only type of mistake possible is 'verbal mistake', this being tacitly assimilated to slips of the tongue. Now if I look at a tulip I may (1) call it a daffodil meaning to call it a tulip, (2) call it a daffodil because I am mistaken about what daffodils are and think flowers of this kind are called daffodils, or (3) call it a daffodil because I think it is a daffodil, even though I know quite well what daffodils, and tulips, are and are like. The first mistake is a slip of the tongue; the second is a verbal mistake of a very different kind, and might be called a 'vocabulary mistake'; the third is a straight-forward mistake of fact. The differences between these three do not concern us here; the point is that we can make mistakes of all three kinds about our sense data. Take the colours we immediately perceive. I may see an expanse of green and (1) call it blue when I meant to say green, (2) call it blue because I am muddled about colour terms and think this colour is called blue, or (3) call it blue because I think it is blue, even though I know quite well that blue is the colour of the deep sea and the summer sky and that green is the colour of grass and cabbages. Philosophers tend to take it for granted that we cannot be mistaken about the colours we see[2] but I, for one, am continually confusing green and blue, not to mention such more subtle colours as buff and stone, fawn and khaki, mauve and maroon, crimson and scarlet, magenta and vermilion.[3] It is comparatively easy to mistake these colours for one another, because one is inattentive, in a hurry, or, as I am, simply bad at distinguishing colours, just as I am bad at distinguishing flowers or makes of motor car. It may be said that I only confuse these colours because I do not know the difference. But the fact that I confuse them does not mean that I do not know the difference and, moreover, if I do not know the difference this

---

way of knowing something, with the consequent implication that no mistake is possible as to the character of what we really see or feel' (III p. 68). This was, for him, virtually *the* sense datum fallacy (II).

[1] Hume, I:I, iv, 2.     [2] Cf. Price on the tomato, I, p. 3.
[3] Cf. Austin II, p. 113.

seems to support the view that I can make mistakes. If I cannot learn the difference between, say, fawn and khaki this is, presumably, because it is easy, and *a fortiori* possible, to confuse the two. And as with colour so with the other sense objects. A man may mistake a whistle for a shriek, a pungent smell for a bitter smell, an apple-wine-type taste for a champagne-type taste, and so on.

Now it may be said that these mistakes are possible only because we are inattentive or in a hurry, or because the thing in question is perceived but fleetingly, or something of the sort; these mistakes could not be made if we attended closely to the object in question. The truth behind this is that the qualities which constitute the sense objects are appearance-determined (5.1), which means that what colours, sounds, etc., I perceive is determined by how the things I perceive appear. And if the colour I see is determined by how it looks to me all I have to do to find out what colour it is is, so to speak, to look and see. I may have to do more than this to find out what colour *the object* is, but I am in the best possible position to find out what colour it looks to me, what colour it is that I see, and so what colour the visual sense object, the coloured expanse I immediately perceive, is.

But the fact that I am in the best position to tell something does not mean that I cannot be mistaken about it. I am alone in this room and so I am in the best position to tell how hot it is in here. But I may have flu coming on and, in fact, be quite mistaken about it. Nor does the fact that I am in the best position mean that I am in a perfect position, that no one can correct me. No doubt no one will ever have sufficient evidence to establish the precise temperature in this room at this moment, but this is logically, and even empirically, possible, and if I do go down with 'flu others will have sufficient evidence to conclude that my present judgement is mistaken. So too with my judgements about how things appear, resemblance sense, to me. I am in the best position to tell, but the evidence of other perceivers, and neurologists, psychologists and whathaveyou, might be enough for others to conclude that I am, after all, mistaken. If there is nothing discoverably wrong with my eyesight, and I say that this paper looks pale blue to me when everyone else says it looks pale green to them, the answer must be, surely, that I have mistaken pale green for pale blue. There is the slight possibility that I do in fact perceive it differently from everyone else, but this is so slight as to make it preferable to say that I am mistaken. And I can, of course, correct my own judgements about how things appear. When everyone else says it looks pale green to them I may very well look again and see that I was mistaken, that it looked, resemblance sense, not pale blue but pale green all the time.

Apart from this there seem to be two reasons why it has been taken for granted that we cannot be mistaken about the colours we perceive. First philosophers have tended to concentrate on the boldest and most obvious colours. Perhaps anybody could confuse crimson with scarlet, even when they know the difference, but who could confuse red with blue? This is rather like asking: who could confuse a piece of chalk with a typewriter? Surely only a slip of the tongue, or at best a vocabulary mistake, is possible here? Nevertheless the question of whether factual mistake is or is not possible remains an empirical one, not to be settled *a priori* by the philosopher. And the fact, if it is a fact, that no one in his right mind muddles red and blue does not prove that we cannot misidentify colours, any more than the fact that no one in his right mind muddles chalk and typewriters proves that we cannot misidentify writing instruments.

Second, colours are determined by how they appear in the resemblance sense of that verb (6.3). But although we can be mistaken about how they appear in this sense, i.e. what colours they are, we cannot be mistaken about how they appear in the other sense, the judgement sense. Take Ayer's example:[1] I look at two lines which are nearly identical in length and say, first, 'Line *A* looks the longer', and then, on closer inspection, 'Line *B* looks the longer'. On the one hand we feel inclined to say that I have discovered that my original statement was false, that I was, at first, mistaken as to which line looked the longer. But on the other hand we feel inclined to say that I wasn't, and couldn't have been, mistaken as to which line looked the longer *to me at that time*. These opposite inclinations stem from different senses of 'looks'. If 'looks' is used in the resemblance sense then the second judgement contradicts the first. It means that I was mistaken in thinking that, to put it roughly, what I saw when I looked first at line *A* and then at line *B* resembled what I would see if I looked first at a long line and then at a shorter one. But if 'looks' is used in the judgement sense what matters is not what the lines are like, but what I

[1] VI, pp. 65–6.

make of them, what I take them to be. And if I say that line *A* looks longer in this sense I cannot be mistaken. For although I may be mistaken as to which line is the longer, and as to which line looks, resemblance sense, the longer, I cannot be mistaken as to which line *I think is* the longer. Nor will my second judgement contradict the first; there is no contradiction between 'I now take line *A* to be the longer' and, said at a different time, 'I now take line *B* to be the longer'.[1]

To say that something is e.g. blue is to classify it, to class it along with other things that happen to be blue, such as the deep sea and the summer sky.[2] Clearly it must be possible to make mistakes, to classify it along with the wrong things. However there is one trivial, although it will be important, way in which a man cannot be mistaken about what he immediately perceives. For I cannot, it may be said, be mistaken in thinking that what I perceive is as it is perceived to be. I may make mistakes in my attempts to classify what I perceive, even what I immediately perceive, but that what I perceive is 'like this', that I perceive these particular colours, etc., whatever they may be and whatever I may think they are, is something I cannot be mistaken about. I will put this by saying that a man cannot be mistaken about the 'immediate appearance' of what he perceives. Even so I would not say that a man *knows* the immediate appearance of what he perceives. I would reject this not because I think that what cannot be false, what we cannot be mistaken about, cannot be known, but because a statement, description, account of immediate appearance is compatible with anything. It rules nothing out and so tells us nothing about what a person is perceiving. Thus the 'knowledge' that what I perceive has such-and-such an immediate appearance has no content, and cannot count as knowledge after all. I seem to

---

[1] Quinton I, pp. 34-5, suggests that the incorrigibility of a 'looks' statement, where 'looks' has what I call its judgement sense, is due to its tentativeness. But as Quinton himself shows tentativeness does not rule out mistake. The incorrigibility is due to the fact that the 'looks' statement is about what the perceiver himself thinks.

[2] Ayer (V, p. 117; VI, pp. 63-4) objects that to say that something is blue is not to say that anything else is blue, or even that anything else exists. But we are not saying that 'blue' *means* 'like the deep sea, the summer sky', etc. We are saying, merely that 'blue' means 'like such things as are blue'. And, as it happens (but only as it happens) the deep sea and the summer sky are blue. So to call something blue is, as it happens, to classify it along with the deep sea and the summer sky.

be saying something but I am not; 'About the philosopher there lingers the air of using language, especially if he says "This is wuf", though of course he might just as well say "Wis is thuf" '.[1] What I say, and know, is merely that what I perceive is perceived to be as it is perceived to be, and this tells nobody, not even myself, anything.

### 11.6 SENSE DATA AND WHAT WE TAKE OURSELVES TO PERCEIVE

We have already considered why people should have thought that sense datum knowledge is knowledge that can be derived from the perception of the particular moment. Why have they also thought that it is absolutely certain knowledge? No doubt the notion of 'knowledge by acquaintance' is at work once again, but for our purposes the most important mistake has been the failure to distinguish the two senses of verbs like 'looks' (cf. 6.3). It is only if 'looks' has what I called its judgement sense that we cannot be mistaken about how things look. Moreover, if 'looks' is given its judgement sense, then the perception of the particular moment is sufficient to give us the right to be sure about how things look to us. This suggests that instead of thinking of sense data as what we immediately perceive, we should rather think of them as what we take ourselves to perceive. And in actual fact much of what philosophers have wanted to say about sense data fits with this interpretation.

Thus it is sometimes said that attention to, closer observation of, what we perceive can alter our sense data. If I notice, first, a piece of wood with wires, and then, on closer inspection, realize that it is a mousetrap, it may be said that my sense datum has changed from one of a piece of wood with wires to one of a mousetrap. Yet what I am aware of has not changed; what has changed is not what I perceive but what I take it to be. So, so long as we do not reject Realism, so long as we keep 'sense datum' the theory-neutral term it is meant to be, we can say that attention alters the sense datum only if we identify the sense datum not with what is (immediately) perceived, but with what we take, judge, consider, what we perceive to be. Only if we adopt a percept theory—and probably not even then—will we want to say that what is sensed on the two occasions is qualitatively different.

[1] Wisdom I, p. 164.

Again it has been argued[1] that sense data must appear to be as they are, cannot possess characteristics which they do not appear to have, this being taken to mean that they cannot possess characteristics that the perceiver does not notice. Many[2] have found this too strange to be plausible. I can, of course, fail to notice what precisely something is like (I see the speckled hen but I do not notice exactly how many speckles it has), but this does not mean that what I perceive is, let alone must be, indefinite in this respect (surely the hen must have some definite number of speckles?). There may even be times where I do not notice what what I perceive is (was it a bird or an aeroplane?) but this does not mean that what I perceive is indeterminate (it must have been a bird or a plane or some, definite, thing). But others[3] have been prepared to allow that what is perceived may be indeterminate in these respects. This much is certain. If what is perceived does not have a definite number of speckles then it cannot be the hen which we would all ordinarily, in our Realist fashion, say we saw. This means that so long as 'sense datum' remains a theory-neutral term we cannot allow that the sense datum is indeterminate in this way. The Realist will have to say that the sense datum includes a definite number of speckles, for the hen possesses a definite number of speckles, and the Realist wants to say that the sense datum includes, in the appropriate sense, part of the skin of the hen. Only if we identify the sense datum not with what is perceived but with what we take ourselves to perceive can we allow that the sense datum might be indeterminate in this respect. For there is no oddity in saying that I take myself to be perceiving a speckled hen without taking myself to perceive any definite number of speckles, as it is odd to say that I perceive something which is speckled but does not possess any definite number of speckles.

Suppose we were to accept a percept theory of perception. Would we, even then, want to think of our percepts as indefinite in the ways described? It seems not, unless we were already convinced that sense data, which we now identify with percepts, cannot have unnoticed characteristics. And why should we think this? For there is nothing very strange, except perhaps to a positi-

---

[1] E.g. Ayer IV, § 4, although this claim is less common than its reverse, that sense data must be as they appear to be, cannot appear other than they are. We will consider this latter claim in the next section.
[2] E.g. Price III; Chisholm I.     [3] E.g. Ayer III; Armstrong I.

vist, about the suggestion that my after-image has features which I fail to notice. In fact there is something very strange about the suggestion that because I do not notice its shape around the top it has no shape around the top! It is significant that Ayer[1] implicitly identifies 'Can sense data have characteristics which the percipient does not notice?' with 'Can sense data have characteristics they do not appear to have?' This identification is only possible if 'appears' has its judgement sense, for an object may well appear $x$, in the resemblance sense, without the percipient noticing that it does. Most of the coins I see look in this sense, elliptical, but I seldom notice the fact. But if an object appears $x$, in the judgement sense, then I must notice, or perhaps 'misnotice', that it does. In so far as I judge or am inclined to judge that the pennies are a certain shape I must notice the pennies and their shape. So if we define sense data as the way things appear, and give 'appear' its judgement sense, it follows that sense data cannot have unnoticed characteristics. But this is to identify sense data not with what is perceived, immediately or otherwise, but with what we take, judge, consider ourselves to be perceiving.[2]

Finally, consider once again the privacy of sense data. If sense data are thought of as what we immediately perceive then, as we saw, their privacy has to be specially stipulated, by laying it down that even if what they perceive is qualitatively the same $A$ and $B$ still perceive different sense data, because they are different perceivers. But if we think of sense data as what we take ourselves to per-

[1] III, p. 90.
[2] Price III makes substantially the same point in terms of a distinction between 'looking' and 'seeming'. Ayer's reply, III, § 5, shows even more clearly that he is identifying sense data with how things appear, in the sense of what the perceiver takes them to be. The reply is that Price is claiming, in effect, that it makes good sense to say 'The flower did in fact look sky-blue to me, but I only noticed that it was bluish'. Now the phrase 'to me' almost invariably goes with 'appears', etc., in their judgement senses, (It looks to me = I (am inclined to) judge that it is) and this sentence does not make sense if 'look' has its judgement sense. But Price's argument was that such a sentence makes sense if 'look' has its resemblance sense, and in fact the statement does make sense if the 'to me' is eliminated, or given a sense appropriate to the resemblance sense (e.g. 'from my point of view'; 'from here' might be more idiomatic). Ayer says that the simple 'It did in fact look sky-blue but I only noticed that it was bluish' describes how it would look to a 'standard observer'. He completely ignores the possibility, envisaged by Price, that it describes how it looks (resemblance sense) to this particular observer in his own particular state, cf. 'I didn't notice the colour, but it must have looked green (to me) because everything looks green (to me) through these spectacles'.

ceive their privacy follows automatically. Two people may take themselves to perceive the same things, but the sense datum perceived by $A$ will be what he takes himself, $A$, to perceive, whereas the sense datum perceived by $B$ will be what he takes himself, $B$, to perceive.

So if we think of sense data not as what we immediately perceive, but as what we take ourselves to perceive we will ensure that we cannot be mistaken about our sense data, that perception of a sense datum is sufficient by itself to give us knowledge of that sense datum, that attention can alter the sense data we perceive, that our sense data cannot have features which do not notice, that our sense data can be indeterminate, and that our sense data are private—and yet still retain 'sense datum' as a theory-neutral term. Thus this interpretation fits with a lot of what people have wanted to say about sense data, and, perhaps without their realizing it, it has in fact been adopted by some philosophers.[1] Is this interpretation preferable to my interpretation of sense data as what we immediately perceive? I think not.

First of all, sense data, on this interpretation, can be said to be perceived only in a highly Pickwickian sense of 'perceive' (from which it follows that their privacy is not privacy in the sense of perceptible by one person alone). If I take myself to be perceiving green when in fact I am perceiving blue I can hardly be said to be perceiving green, yet, on this account, the sense datum will be green. Similarly the man who takes a vine to be a snake will, on this account, have a sense datum of a snake, even though he perceives no such thing. Moreover, I think it follows from this that sense data, thus interpreted, cannot be the sorts of things that percept theorists identify as percepts. For percepts are things we perceive; indeed according to the theory they are the only things we perceive. So since 'sense datum' is originally introduced as a theory-neutral equivalent for 'percept', it follows once again that sense data, thus interpreted, are different from sense data as traditionally conceived.[2] Personally I would reject this interpre-

---

[1] E.g. Ayer III, as I have been arguing, and Armstrong I. Similarly Warnock I, ch. 7. His suggested description of a Berkelean idea, 'It seems to me as if I were hearing a sound', is not a description of what I perceive, so much as an account of what I make of what I perceive, what I take it to be. Cf. p. 176 n above.

[2] Cf. also Austin's point, II, p. 113 n: 'To stipulate that a sense datum just is whatever the speaker takes it to be—so that if he *says* something different it

tation of sense data. If we want to talk about what we take ourselves to perceive it is much better to talk about 'perception-that'.

Nevertheless one important point has emerged from all this. Sense datum knowledge is not, as we hoped, knowledge derivable from the perception of the particular moment. Nor is it absolutely certain knowledge. But knowledge of what we take ourselves to perceive—let us call it 'knowledge of perception-that'—is both of these things. The fact that I take myself to perceive something does, by itself, give me the right to be sure that I take myself to perceive that thing. And if I think that I perceive something, I cannot be mistaken in thinking that that is what I think it is. Notice that it is not because I cannot be mistaken that the knowledge is derivable from the perception of that particular moment. Rather the fact that a person cannot, logically cannot, take himself to be perceiving something without knowing that he does, means that if I perceive something in the strong sense (2.3) I must, *ipso facto*, have the right to be sure that I take myself to be perceiving whatever I do. Nevertheless the incorrigibility of this knowledge is connected with the fact that it is knowledge derivable from the perception of the particular moment—it is the same feature as makes it incorrigible as also makes it derivable from that perception.

So it looks as though it will be knowledge of perception-that, and not sense datum knowledge, which will provide a foundation for empirical knowledge. The plain fact is that, for various reasons, philosophers have confused or failed to distinguish the two, thinking that what is true of the former is also true of the latter. The most important reasons for this seem to be (1) the failure to distinguish between the two senses of 'appears', 'looks', etc.; (2) the theory of knowledge by acquaintance; (3) the quest for the incorrigible, which tempts us to slide from the fact that we are unlikely, particularly if we are careful, to make mistakes over e.g. the colours we see, to the conclusion that we cannot make mistakes over them; (4) the failure to distinguish between what does not go beyond the perception of the particular moment, and what is derivable from the perception of the particular moment.

must be a different sense datum—amounts to making non-mendacious sense datum statements true by *fiat*; and if so, how could sense data be, as they are also meant to be, non-linguistic entities *of* which we are aware, *to* which we refer, that against which the factual truth of all empirical statements is ultimately to be tested?'

## 11.7 SENSE DATA AND APPEARANCES

While we are on the topic of the confusions engendered by these words 'appears', 'looks', 'sounds', etc., we might say something more about the interpretation of sense data as the appearances which things present to us. We have already (11.2) seen that there are difficulties and obscurities in defining, e.g. visual sense data as 'The way the things we see look to us'. Perhaps these difficulties can be overcome, but I think there are other considerations which show that this interpretation is definitely unacceptable.

But first, some preliminary points. Identifying sense data with the appearances things present to us will not mean that we cannot be mistaken about our sense data. If, as is likely, I do not notice that the tilted penny looks elliptical I may not realize that it is tilted, and so, when asked if it looked elliptical, say that it did not. And even if I do realize that it is elliptical in appearance I may make a mistake about the precise elliptical shape it appeared. If I am shown a series of ellipses and asked which has the shape the penny appeared to have I might very well pick out the wrong one. And we have seen that I can be mistaken about what colour a thing looks. Absolute certainty is ensured only if 'looks' and its associates are given not their resemblance but their judgement senses. Second, if we identify sense data with the appearances things present to us their privacy will have to be specially stipulated. If two people look, one after another, at the same unchanged objects from the same point of view we would naturally say those objects present the same appearance to them both. If we want sense data, conceived as appearances, to be private we will have to lay it down that it follows logically from the fact that they are different perceivers that they perceive different sense data. Finally, with non-appearance-determined qualities it is possible for an object to appear both $x$ and $y$, where $x$ and $y$ are incompatible properties. The tilted penny looks round and also looks elliptical. So it will have to be specified which of such incompatible apparent qualities of the object is to be the actual property of the sense datum. Many, over-impressed by the Berkelean arguments against the possibility of perceiving distance, a line turned endwise to the eye, suggest that since third-dimensional characteristics cannot be perceived, but have to be inferred or assumed or constructed or taken for granted on the basis of past experience, the sense datum shape will be the shape what is seen appears when it is construed two-dimensionally. On this account the sense datum of the tilted penny will be elliptical. But others, in particular Price,[1] have argued, quite correctly to my mind, that what is seen can and does exhibit a certain visual depth which is not a construction from past experience or anything of that sort, and that in so far as what is seen does exhibit visual depth the sense datum shape will be the shape that what is seen appears to have when construed three-dimensionally. On this account the sense datum of the tilted penny will possess the shape of a tilted circle, whereas the sense datum of a distant mountain will be flat (at that distance the phenomenon of visual depth ceases to apply). Size is even more difficult to deal with. The moon looks as big as a sixpence, or a threepence, or a penny, or something thousands of miles across, depending, so to speak, on how you look at it. We will need to stipulate, arbitrarily, some criterion such that if, on this criterion, $x$ looks as big as $y$, then the sense datum of $x$ is to be said to be $y$-sized. An appropriate criterion might be how big things look when held at arm's length: if the moon looks as big as a sixpence held at arm's length, the sense datum of the moon will be said to be as big as a sixpence.

However, none of these considerations rule out the identification of sense data with the appearances things present to us. I think the next two do show that this identification is unacceptable. First, it is usually considered an essential feature of sense data that we perceive them whenever we perceive. But although it may be true that appearances are involved in all perception, in that whenever we perceive or even merely sense something it can be said to present an appearance to us, it is perfectly possible for a person to perceive something without noticing, and hence without perceiving in any full or adequate sense, the appearance it presents to him. I may see the tilted penny without noticing that it looks elliptical; I may see something which looks brown without noticing that it looks brown, either because I am inattentive, or because it went by too fast for me to notice. The idea that if one perceives $x$ one must know how it appears is due, once again, to the failure to distinguish the two senses of 'appears', for if 'appears' is used in its judgement sense I cannot perceive something without knowing how it appears, what I take it or am inclined to take it to be.

[1] E.g. IV.

## SENSE DATA

Secondly, this interpretation of sense data prevents 'sense datum' from being a theory-neutral term. When I look at the distant mountain it looks flat. The phenomenon of visual depth is missing so even those who follow Price in allowing visual sense data to have three-dimensional characteristics will want to say that the appearance is flat. And if the appearance is the sense datum it follows that the sense datum if flat. But in that case the coloured expanse I immediately see cannot be the surface of the mountain, as the Realist wants to say it is, for the surface of the mountain is certainly not flat. If 'sense datum' is to be a theory-neutral term we cannot accept the identification of sense data with the appearances things present to us as we perceive them, the claim that sense data *are* what the objects *appear* to be. For if we say that the sense datum is as what is perceived appears to be, we have, in so far as what is perceived appears from what it really is, to reject the Realist's claim that sense data include parts or aspects of external objects.

There are, however, important connections between describing what is immediately perceived, the sense datum, and describing the appearances things present to us. No doubt it is these connections which hide the fact that 'appearances' and sense data, though related, are different. Another, less direct, source of the identification of sense data with appearances is the tendency to confuse what is immediately perceived with what we take ourselves to perceive, together with the failure to distinguish the two senses of 'appear', so that what we take ourselves to perceive, how things appear to us in one sense, is identified with how they appear to us in the other sense.

This failure to distinguish the two sense of 'appears', and related verbs, has been seen to be the source of many mistakes and confusions. It may help to list them here:

The failure to distinguish how things appear to us, in the sense of the appearances they present to us, from how they appear to us, in the sense of what we take them to be, can lead us:

(1) to think that since we cannot be mistaken about how things appear to us in the second sense we cannot be mistaken about how they appear to us in the first sense;

(2) and even to identify the appearances things present to us with what we take ourselves to perceive;

(3) and to think that if the thing appears, resemblance sense,

other than it is we must take it to be other than it is, this in turn tending to make us identify appearance and illusion with mistake, delusion and non-veridical perception generally;

(4) and eventually, via (2) and various other factors which lead us to identify what we take ourselves to perceive with what we immediately perceive (cf. 11.6), to identify the appearances things present with what is immediately perceived.

We must distinguish these three things—what we immediately perceive, the appearance things present to us, and what we take what we perceive to be—and it seems that it is with the first of the that sense data are, most happily, to be identified.

Nevertheless, even if sense data are not to be identified with the appearances things present to perceivers there will be definite connections between the sense datum I immediately perceive and the appearance of what I perceive. First, so far as the appearance-determined qualities go, the quality which an object appears, resemblance sense, to have will also be the quality ascribed to the sense datum. If what I see looks red then the sense datum I immediately see will be red. Nor is it an accident that it was just these appearance-determined qualities which we fastened on in our account of the various sense objects. For it is because what quality we perceive is, in these cases, determined by how things appear, resemblance sense, that the statement that we are perceiving an instance of this quality does not go beyond the perception of the particular moment.

As for non-appearance-determined qualities: We have seen that we cannot, without rejecting Realism, say that when I see the tilted penny I immediately perceive an elliptical expanse of brown. For according to the Realist the brown expanse I immediately perceive is the surface of the penny, and the surface of the penny is not elliptical. Nor, for that matter, can we say that the visual sense object is circular, although tilted in the third dimension, as according to the Realist it is. For to say this is to say something about how it would look from other angles, to go beyond the perception of the particular moment, and so go beyond a sense datum statement. The most we can say, in a sense datum statement, is not that the visual sense object *is* circular but tilted, but that it *looks* circular but tilted. So far as the non-appearance-determined qualities are concerned, sense datum statements are restricted to saying what they appear to be, and cannot say what they are. In this respect too the sense datum statement resembles an account of the appearance of what is perceived. But there is this difference: the sense datum statement will be that the sense datum *looks*, e.g. elliptical, whereas the description of the appearance of what is perceived will be that the appearance *is* elliptical. The two accounts must be different; it would hardly make sense to say that the appearance of what is perceived appears elliptical!

A consequence of this is that the only 'identification' (cf. 6.2) legitimate in a sense datum statement is an identification of the appearance-determined qualities that happen to be perceived. I mentioned (11.2) White's suggestion that a sense datum statement is a description, as opposed to an identification, of what is perceived, and I argued that what is out of place in a sense datum statement is not so much the identification of what is perceived as any assertion that what is perceived really exists. Even so a case could be made out for allowing White's 'descriptions' as sense datum statements, for if Macbeth were to say 'A dagger' (construed as description rather than an identification) or 'A long thin tapering shining object' these descriptions of what is perceived could be taken as not going beyond what is perceived at the particular moment. This would be tantamount to taking them to assert that what is perceived is like a dagger, or like a long thin tapering shining object, without committing the speaker to any claim about what, if anything, is really there or what it is really like, i.e. as equivalent to the explicit sense datum statement that what is perceived is a bright coloured expanse of such and such an apparent shape. But although White's 'descriptions', as thus interpreted, are equivalent to sense datum statements I prefer my own account, on the grounds that it is more precise (White's 'descriptions' can, after all, be interpreted in other ways in which they would not be equivalent to sense datum statements), brings out more clearly how sense data are to be thought of (as consisting of sense objects, etc.) and does not carry the false implication that accounts of what is immediately perceived cannot, in any way at all, identify what is perceived.

## 11.8 THE IMPORTANCE OF SENSE DATA

Sense data and sense datum terminologies have become something of a philosophical Aunt Sally. This seems to be due to three things: the failure to distinguish sense data from percepts, the dislike of the notion of knowledge by acquaintance (and the claim for incorrigibility which goes with it), and a distrust of traditional talk about inferring from sense data, as if we spent our days carefully noting and interpreting our sense data. But it seems to me that a sense and a use can be given to the notion of a sense datum, and in this section I want to bring together some of the ways in which it might be helpful.

The first important point about sense data is that they are what we immediately perceive, much as it is a philosophically important fact about events that they occur. These are, of course, tautologies, but not pointless tautologies, and they are certainly more than arbitary terminological points—saying that all perception involves perception of sense data is not like saying that all vision involves semi-vision, 'semi-vision' being defined as visual perception of the top half of the visual field. The importance of the analytic truth that all perception involves perception of sense data can be brought out by considering a rather different analysis of perception from that considered in Chapter 2, one which does not conflict with, but rather cuts across, that earlier account. We can say that perception, in the weak sense, of $x$ consists in (a) the immediate perception of a certain sense datum $s$, and (b) the fact that the sense datum $s$ is a sense datum of $x$. Similarly perception, in the strong sense, of $x$ consists in (a) the immediate perception of a sense datum $s$, and (b) realizing that $s$ is a sense datum of $x$. The precise relationship expressed by the 'of' is disputed by the different theories of perception, but for a Realist, to say that $s$ is a sense datum of $x$ will mean that $s$ includes parts or aspects of $x$. How is this analysis illuminating? Suppose I see a house. It is not, of course, necessary that I perceive every smallest part of the house, for this is something I may never do with any house. Nor is there one particular part of the house that I have to perceive in order to be said to perceive the house. It may be that there are some parts such that perception of them is not sufficient to count as perceiving the house, but nevertheless there are several different parts such that perception of them is sufficient to count as perceiving the house. To put it paradoxically: different people can perceive different things (different parts of the house) and yet perceive the same thing (the same house). And, on the other hand, what people perceive can be qualitatively the same without it being true that they perceive the same thing, as when they see different houses in the same terrace. Our perception of the house can be divided into

(a) our awareness of whatever it is, and (b) the fact that it is (part of) a house. This is brought out by our analysis of perception in terms of the perception of sense data.

The analysis shows, in a way which for perfectly good reasons our ordinary way of describing what we perceive does not, how our perception of things is limited. There are many things we can and do say about what we perceive, but most of these go beyond what we perceive. If I am to restrict myself to an account of what I an aware of through my senses, without relying on any other knowledge or beliefs about what I perceive, then I must restrict myself to a sense datum statement. It isn't just that all perception involves perception of sense data. It is that we perceive things *only in so far as we perceive sense data of them.* 'Strictly speaking' we perceive nothing but sense data. The notion of a sense datum thus brings out more clearly what precisely it is to perceive something.

The notion of a sense datum might also be of use in a phenomenological examination of what it is to perceive, and what sorts of things we perceive via the various senses. Not, I think, that this would be very important use for the notion of a sense datum. Perhaps the same job could be done more straight-forwardly by means of the notion of appearance.

Finally, since the sense datum is what we perceive of what we perceive as we perceive it to be, and since we perceive things only in so far as we perceive sense data of them, it will be our sense data which, in the last analysis, explain our knowledge and conception of the world around us. That is, it is because we perceive the sense data we do that we have the knowledge and conception of the world we do. Our sense data are, in an important sense, our basis and evidence for those facts and beliefs about the external world which we hold to be true (which is not to say that our knowledge and conception of the external world is inferred from the sense data we perceive). Because of this it is inevitable that sense datum knowledge should have an important place in any theoretical, as opposed to historical or psychological, account of how perception can provide us with knowledge of the external world. If perception does provide us with this knowledge it will come from what we perceive on particular occasions, and in order to discover what we perceive on particular occasions and what knowledge this can provide us with, we will have to consider sense data and sense datum knowledge.

# CHAPTER 12
# OUR KNOWLEDGE OF EXTERNAL EXISTENCE

## 12.1 THE STRUCTURE OF SENSE KNOWLEDGE

We were looking for a foundation of empirical knowledge which would be both knowledge derivable from the perception of the particular moment, and knowledge about which we could not possibly be mistaken. Unfortunately sense datum knowledge turned out to be neither of these things. Nevertheless we saw that those who think that sense datum knowledge provides a foundation in both these respects, have tended to confuse sense data, what we immediately perceive, with what we take ourselves to perceive. And the interesting thing is that knowledge of what we take ourselves to perceive, 'knowledge of perception-that', is both derivable from the perception of the particular moment, and such that we cannot be mistaken about it. It seems that our account of how we know that external objects exist should begin from knowledge of perception-that.

If I am to know that what I perceive does really exist, I must know that I make no mistakes of perception-that, that I do perceive what I think I do, and that what is thus perceived does really exist. So we might say that I have to pass first from knowledge of what I take myself to perceive to knowledge of what I actually do perceive, i.e. from knowledge of perception-that to sense datum knowledge, and then from knowledge that I perceive something to knowledge that that something does really exist, i.e. from sense datum knowledge to what I will call 'knowledge of reality'. In this way we can come to think of a structure or hierarchy of sense-knowledge, where each step or level is thought of as more basic and fundamental than the one above, and where, it is hoped, each level can be built up from the one below:

Knowledge of Reality, knowledge that the things perceived do exist.

Sense Datum Knowledge, knowledge that various things are perceived.

Knowledge of Perception-that, knowledge that I take myself to perceive various things.

OUR KNOWLEDGE OF EXTERNAL EXISTENCE 205

This picture of our empirical knowledge is old-fashioned, and no doubt misleading, but don't let's worry about that. The question is whether it reveals any important truths about the nature and source of our knowledge of the external world. I think that it does, but first the list calls for three comments.

First, it may seem strange that knowledge of perception-that should rank lower than sense datum knowledge, for more often than not my judgements as to what I perceive, my perception-that, go much further than a mere account of what I immediately perceive. I may, for example, take myself to be perceiving Goya's portrait of the Duke of Wellington, and to think this is, obviously, to go beyond what I immediately perceive. However the question is not whether perception-that goes beyond immediate perception, but whether knowledge of perception-that goes beyond knowledge of immediate perception, sense datum knowledge. The knowledge that I am, in fact, perceiving Goya's portrait of the Duke of Wellington does go beyond sense datum knowledge, but it is not knowledge of perception-that, it is knowledge of reality. The knowledge of perception-that is the knowledge not that this is what I *am* perceiving but that this is what, rightly or wrongly, I *take myself to be* perceiving. And we can see that this latter is required for sense datum knowledge, or any other knowledge about what I perceive, although not vice versa.

Next, there may seem to be a need for a category of knowledge which stands between knowledge of reality and sense datum knowledge, viz. knowledge of what we perceive, whether it really exists or not, which is not restricted to knowledge of what we immediately perceive. But what would be an example of such knowledge? We might say that my knowledge that I perceive a rose, where this does not include any suggestion or implication that such a rose does or does not really exist, is an example of such 'knowledge of perception', as it might be called, which goes beyond knowledge of immediate perception. However to say that what I perceive *is* a rose, is to 'identify' (6.2) what I perceive as a rose, and so is to say that it is an external object, for roses are physical objects and physical objects are external objects. So my knowledge that what I perceive is a rose is knowledge of reality. If I am to exclude this implication I must say only that what I perceive is *like* a rose, i.e. 'describe' it as a rose. And we have seen (11.7) that such a 'description', as opposed to an 'identification', of what is perceived is tantamount to a description of the sense datum that is immediately perceived—if the description of what I perceive as 'rose-like' rules out any suggestions or implications about what is really there it is equivalent to the sense datum statement that I perceive various coloured expanses of various apparent shapes, etc. The moral is that any account of what we perceive that does not involve any claim about what really exists will be equivalent to a sense datum statement, or at most to a set of sense datum statements (as when we want to describe what several people perceive at different times without wanting to say or imply that what is perceived does or does not really exist).

Finally, one of the main objections to this 'ladder' picture of our knowledge is that it completely distorts the way in which, in actual fact, we do come to know such things as that there is a spider on the wall. It is ridiculous to suggest that I go through some chain of reasoning such as 'I take myself to be perceiving a spider on the wall, so I am immediately perceiving a spider-on-the-wall-like sense datum, so there is a spider on the wall in front of me'. However no claim is being made for the role of knowledge of perception-that or sense datum knowledge in our ordinary day-to-day acquisition of knowledge about the world in which we live. Epistemology is not concerned with what goes on when we learn various things, nor with what, in a particular case, provides us with knowledge of some particular fact. It is concerned with the theory of knowledge, with what knowledge is, what is acceptable as knowledge, what types of knowledge underlie other types, what kinds of things are in general acceptable as providing the right to be sure, and so on.

Now I think we can see that this picture of a hierarchy of sense knowledge is, theoretically speaking, valid and informative, at least from an Empiricist point of view. For, given Empiricism, each type of sense knowledge is 'epistemologically prior' to the type listed above it, and 'based on' the type listed below. What I mean is this:

I will say that $X$ knowledge is *based on* $Y$ knowledge if for any piece of $X$ knowledge there is always some piece of $Y$ knowledge which was necessary for the having of that $X$ knowledge, not in the sense that the $X$ knowledge is impossible without the $Y$ knowledge, but in the sense that the particular person would not have the $X$ knowledge if he did not have, or had not had, the $Y$ knowledge. Thus my knowledge that Brutus killed Caesar is based on my knowledge that my history teacher told me so, in that I wouldn't know that Brutus killed Caesar unless I knew, or had known, that my history teacher had told me so. Notice first that my knowledge of what the teacher said is not sufficient for my knowledge that Brutus killed Caesar; rather it is necessary. Notice second that it is necessary not in the sense that I *couldn't* know this unless I knew what the teacher said, but in the sense that I *wouldn't* know it unless I knew what the teacher said.[1] And notice third that to prove the general thesis that knowledge of one kind is based on another kind more is required than citing an instance in which it is.

I will say that $X$ knowledge is *epistemologically prior* to $Y$ knowledge if $Y$ knowledge is impossible without (some piece of) $X$ knowledge. Thus knowledge of things I do perceive is epistemologically prior to knowledge of things I do not perceive if it is the case, as it seems to be, that I couldn't know anything about what I have not perceived if I did not know something about things I have perceived. This differs in two important respects from talking about one sort of knowledge as based on another: first the claim is not that there is some specific piece of $X$ knowledge which is necessary for each specific piece of $Y$ knowledge but only that there must be some $X$ knowledge—which may differ from person to person—if there is to be any $Y$ knowledge; and second the claim is that the $X$ knowledge is necessary for the $Y$ knowledge, not just in the sense that a particular person *wouldn't* have the $Y$ knowledge unless he had some $X$ knowledge, but in the sense that he *couldn't*.[2] Nevertheless the two are related in that epistemo-

---

[1] Cf. the account of knowledge 'based on' perception, sense knowledge, in 10.6.

[2] Is this 'can' logical or empirical? If empirical we seem to be trespassing into psychology and trying to establish *a priori* what is or is not within human capabilities. But if the 'can' is logical we seem to be leaving epistemology for logic since, presumably, the only way in which one sort of knowledge would be logically impossible without another would be where the one entails the other. And apart from reservations about the notion of knowledge entailing knowledge (cf. p. 210 n) there is the point that the logical 'can' seems to let in such possibilities as innate knowledge implanted in our minds by a beneficent God! I think we have to say that the 'can' is logical, but with an empirical qualification, thus placing talk about epistemological priority somewhere between talk about how, in fact, human beings acquire knowledge, and talk about what logical connections, if any, hold between different types of knowledge. I will say that $X$ knowledge is epistemologically prior to $Y$ knowledge if $Y$ knowledge is logi-

logical priority involves being the basis of—if $X$ knowledge is epistemologically prior to $Y$ knowledge it follows that $Y$ knowledge must be based on $X$ knowledge. So the easiest way of showing that $Y$ knowledge is based on $X$ knowledge is to show that it must be, that $X$ knowledge is epistemologically prior to $Y$ knowledge.

The suggestion is, then, that, for the Empiricist at any rate, sense datum knowledge is epistemologically prior to knowledge of reality and knowledge of perception-that is, in turn, epistemologically prior to sense datum knowledge; or, to put it the other way around, that sense datum knowledge must be based on knowledge of perception-that and knowledge of reality, in turn, must be based on sense datum knowledge. To test this suggestion we ask, first, whether a man could know that what he perceives does really exist without knowing that he was perceiving a certain sense datum. It may seem that this is possible, for we seldom if ever realize that we are perceiving sense data, but this reply rests on certain misunderstanding. For a person to know a certain fact it is not necessary that he explicitly consider this fact, let alone realize that or consider whether he knows it or not. At 5.30 last night I knew[1] that my name was Don Locke although nothing, at that time, was further from my thoughts. Moreover it is possible for a person to know about $x$s even when he doesn't know what $x$s are, in the sense that he doesn't know the application of the name '$x$', so long as he knows what $x$s are, in the sense that he is familiar with the things which are, as it happens, called $x$s. We are all familiar with expanses of colour, and know when we perceive them, and so we

cally impossible without (some piece of) $X$ knowledge, in the sense that *given that we have the ways of knowing we do and no others* we could never have the right to be sure about, and hence could never know, anything in the $Y$ range unless we already knew something in the $X$ range. If $X$ knowledge is epistemologically prior to $Y$ knowledge then the self-contradiction will be not '$A$ has a piece of $Y$ knowledge but no $X$ knowledge' but '$A$ has a piece of $Y$ knowledge and no $X$ knowledge, and yet has no ways of knowing other than those possessed by normal human beings'. This definition has the merit of turning our enquiries towards the crucial question of what ways of knowing are, in fact, allowed by our accepted concept of knowledge. The question whether I could come to know that Brutus killed Caesar without knowing anything about what I have perceived is not a psychological question about how people learn such facts, but the epistemological question of whether or not we would be prepared to allow that someone had the right to be sure of this without knowing anything about what he has perceived.

[1] This does not mean that knowing is a datable occurrence, but only that the answer to 'Does he know what his name is?', asked at 5.30 last night, was 'Yes'.

are all familiar with visual sense data, and know when we perceive them, even though we may not know how to use, much less explain, this term 'sense datum'.

So our question is: can a person have the right to be sure that what he perceives really exists, without having the right to be sure that he perceives certain sense data, i.e. sees various expanses which are coloured, hears various sounds, smells various smells, etc., where these expanses, sounds, smells, etc., lie within the range of what he immediately perceives. And, given Empiricism, the answer must be that he cannot. I know that I am perceiving a typewriter, and one that really exists, but surely I could not know this, at least not from my perception, if I did not know that I am perceiving various expanses of colour, which are, in fact, parts of the surface of the typewriter? This is not, to repeat, to say that I explicitly realize that I am perceiving those expanses, let alone that I first note that I perceive them and then infer, conclude or come to the opinion that there is a real typewriter there. Nor is it to say that I notice or know the precise nature of those expanses which, as it happens, make up my visual sense datum. Nor is it to say that I realize or know that those expanses comprise a sense datum, or that 'sense datum' is a term which can be used to refer to those expanses in so far as I perceive them. All it is to say is that just as a person cannot perceive a really existing item without perceiving a sense datum, so he cannot know that he perceives a really existing item, know that what he perceives really exists, without knowing that he perceives certain features of that item, those features which figure in his sense datum of that item. For in order for his perception to inform him that the item really exists he has knowingly to perceive something, and in order to know that he perceives that something he has to know that he perceives certain parts or aspects of that thing, parts or aspects which figure in his sense datum.

Next, is knowledge of perception-that epistemologically prior to sense datum knowledge? Can a person know something about what he perceives, or for that matter anything about the world around him, unless he knows what he takes himself to perceive? It seems not, for a person cannot take himself to perceive various things without knowing that he takes himself to perceive them and, surely, a person cannot know anything about what he perceives unless he takes himself to perceive something.

We see, then, that for the Empiricist there is this much truth in our hierarchy of sense knowledge, that each type of knowledge listed is epistemologically prior to, and hence the basis of, knowledge of the type listed above it. Indeed this seems to be a consequence of the Empiricist thesis itself—what we have done is bring out how the Empiricist might think of one sort of knowledge as built on, and out of, another. We must now look more closely at the nature of this building.

## 12.2 REDUCTIONISM AND CONSTRUCTIVISM

The most common way in which philosophers try to build up one type of knowledge from another is by means of what I will call *logical construction* or, looked at from the other end, *logical reduction*. The claim that one type of knowledge is constructible out of or reducible to another involves, I think, two things: first that knowledge at the higher level is entailed by, logically deducible from, knowledge at the lower level,[1] and second that the knowledge at the lower level is in some sense simpler and more basic than the knowledge it entails. Presumably this simplicity consists in the fact that it takes more than one 'piece of knowledge', knowledge of more than one fact, to entail the knowledge at the higher level. Thus an example of a reductionist analysis might be the reduction of the knowledge that it has rained on the last three days into the separate pieces of knowledge that it rained on such and such an occasion, that it rained on such and such another occasion, and that it rained on such and such a third occasion, that these

[1] There is a difficulty in talk of knowledge entailing knowledge. For that $A$ entails $B$ does not mean that knowledge of $A$ entails knowledge of $B$. 'Father' may entail 'male', but it is possible to know that $x$ is a father without knowing that he is male—an a-sexual Martian may know many facts about fathers (that they are heads of households, parents, opposites of mothers, etc.) without knowing that they are male, perhaps even without knowing what males are. For knowledge of $A$ to entail knowledge of $B$ it is necessary not only that $A$ entail $B$ but that the person know that it does; it is only together with the knowledge that being a father involves being male that the knowledge that $x$ is a father entails the knowledge that he is male. However reductionists are usually interested in the relationships between the facts as much as the relationships between the knowledge of the facts, so let us, for convenience, ignore this point and say that knowledge of $A$ is reducible to knowledge of $B$ where $B$ entails $A$, even though it may still be possible to know $B$ without knowing $A$. When we talk about logical constructions and reductions we are concerned with the connections between what is known, rather than with the connections between the knowing of it.

occasions occurred on consecutive days, and that the last occasion occurred yesterday.

The reductionist/constructivist motive has played an important part in Empiricism and traditional epistemology generally, although there is less interest now in that old topic 'the construction of an external world' than there once was. Perhaps this is because the considerations which underlie such constructions are somewhat discredited, in particular that pious philosophical belief that there are always basic, ultimate, simple elements out of which everything of a certain sort is constructed and to which everything of that sort can, in the last analysis, be reduced. This usually goes with the hope, in matters epistemological, that it is by getting down to these basic, simple, ultimate elements that we can attain absolute certainty and so exorcize, at last, the bogey of scepticism which has, consciously or unconsciously, haunted philosophers since Descartes. But, discredited or not, an epistemological construction of the external world, i.e. a logical construction of our knowledge of the external world, would yield three important results.

First, if we begin with what perception by itself is sufficient to give us the right to be sure of, and then carefully build up step by step until we have covered all our knowledge of the external world, we will have provided just the proof of Empiricism we are looking for. That is, we will have shown how all our knowledge of the external world is derivable from perception alone, and so have offered a substantial reason for thinking that all such knowledge is derived from perception alone, that perception is the sole source of such knowledge. We will thus have answered the fundamental epistemological question of how, as regards our knowledge of the external world, we know what we know. We will have shown what gives us the right to be sure, and how it does so.

Second, if the basic atoms from which this knowledge is constructed are such that we cannot be mistaken about them, we will be provided not only with a final answer to the regressive sceptic, we will also have an answer to a dogmatic scepticism. For a logical construction of knowledge from unquestionable premises will not only provide an unquestionable proof that we do know, it will also show that all the constructed knowledge is unquestionable too, is such that we cannot be mistaken about it and so fits with even the sceptic's specially high standards for knowledge.

Third, if we think of these basic atoms as not just epistemological units but also as some sort of physical or metaphysical entity, e.g. percepts, the epistemological construction of our knowledge can be duplicated by a metaphysical construction of the world itself. Once it is accepted that all our perception is perception of percepts, and that all our knowledge of the world is built up out of knowledge of such percepts, it will be natural to conclude that the world itself is built up out of percepts in just the way that the knowledge is built up out of knowledge of percepts. However a demonstration that our knowledge of the world can be constructed from sense datum knowledge would not be a proof that the world itself is constructed of sense datum entities. That follows only if we adopt the extreme positivistic thesis that things exist and are constituted only in relation to our knowledge of them, in the sense that if our knowledge of $x$ is built up out of knowledge of $y$ and $z$, then $x$ itself must be built up from $y$ and $z$.

Can this construction/reduction be carried out? Can we arrive at our knowledge of the real existence of what we perceive via a series of entailments commencing from knowledge which is absolutely certain and derived from the perception of the particular moment? Given our hierarchy of sense knowledge this becomes two questions: can we reduce knowledge of reality to sense datum knowledge? and can we reduce sense datum knowledge to knowledge of perception-that?

To begin with the former: my knowledge that there is a typewriter here in front of me is knowledge of reality, but it is not entailed by my knowledge that I am now perceiving certain typewriter-like expanses of colour, not even together with the knowledge that I have perceived various similar sense data in this place on so many occasions in the past, plus the knowledge that I have, on so many occasions, perceived sense data as of people acting and talking as if they, too, perceived typewriter-like sense data. For it is logically, perhaps even empirically, possible for a person to suffer a pervasive hallucination such that although he perceives just such sense data as I have of this typewriter no such object exists. This is even more obvious if we take my knowledge that this drawing, which I have just drawn, really exists. For it may well be that no one else ever sees this drawing, and that I never see it again, and although it would be silly to say that I do not know that it exists it certainly doesn't follow from the fact that I perceive these sense data that it does exist. It is not at all uncommon for people to perceive just such sense data and yet not perceive something that really exists. Of course if a large number of people perceive sense data of the appropriate sort it cannot be said that what they perceive is a hallucination; the fact that so many people have over so many centuries and in so many different ways perceived sense data as of this island we call Great Britain does entail that such an island exists. But two points need to be noticed about this case. First, this reduction of the knowledge of reality that Britain exists to sense datum knowledge is a reduction to the sense datum knowledge of *different* people, it is not a reduction of *my* knowledge of reality to *my* sense datum knowledge. It is not usually clear whether such a reduction is supposed to be to the knowledge of the individual perceiver or not, but in so far as the aim is to provide an answer to the sceptic I think it will have to be. For if the reduction brings in reference to other people's knowledge the sceptic will want to know what gives me the right to be sure that other people have this knowledge. There is no sense datum knowledge which I do or could have, from which it follows logically that other people perceive those sense data, have that sense datum knowledge. And second, although it may be possible in some cases to reduce knowledge of reality to the sense datum knowledge of different perceivers this does not help the reductionist so long as there is other knowledge which it is not possible to reduce in this way. For the claim is that *all* our knowledge of the external world is so reducible, and, as the examples of the typewriter and the drawing show, this claim is mistaken.

At this point we may be told that we need to interpret the reduction in a different way. It may be said that no one claims that all our knowledge of the external world can be reduced to sense datum knowledge, and so to knowledge of perception-that, which people actually possess. For that would mean that all our knowledge of the external world is, like the knowledge of perception-that from which it can be deduced, absolutely certain, such that we cannot be mistaken about it. And that, clearly, is false. What is maintained, the suggestion is, is not that we *do* have absolutely certain knowledge from which follows, logically, all our knowledge, but that for any piece of knowledge we *could* provide ourselves with absolutely certain knowledge from which that knowledge would follow. The answer to the sceptic is not that all our knowledge is absolutely certain but that, if it is knowledge, we could, by taking the appropriate steps, make it so. I may not perceive sufficient sense data for it to follow logically that there really is a typewriter or a drawing here, but I, or at any rate people generally, could perceive sufficient sense data for this to follow. With this suggestion reductionism joins hands with Phenomenalism. Just as the Phenomenalist claims not that facts about external objects are equivalent to facts about sense data which have been perceived, but that they are equivalent to facts about sense data that could be perceived, so the present suggestion is that knowledge about external objects is reducible to and entailed not by knowledge of sense data that have actually been perceived, but by knowledge of sense data that could be perceived. The answer is much the same in both cases. This 'potential reduction', as we might call it, of knowledge of reality to sense datum knowledge can only be achieved if we refer to the sense data, and the sense datum knowledge, of more than one perceiver. And even though this potential reduction to the sense datum knowledge of various perceivers can be carried out, to say that it can, i.e. to say that the sense datum knowledge in question could be obtained, presupposes that the object in question does really exist, just as the Phenomenalist's notion of a possible sense datum presupposes the notion of real existence. And this deprives us of our reply to the sceptic, for if we say that sense data, and sense datum knowledge, could be obtained from which it followed, logically, that there is a drawing here, the sceptic will want to know how we know that they could be obtained. The honest answer is that in saying that they could be obtained we are presupposing that the drawing does really exist.

We can see, then, that a reduction of a sort, although not, it appears, of the sort hoped for, can be provided from knowledge of reality to sense datum knowledge. Even so the reduction of our knowledge of the external world to absolutely-certain, derived-from-the-perception-of-the-particular-moment knowledge of perception-that does not succeed, for it fails completely at the next step, in the attempt to reduce sense datum knowledge to knowledge of perception-that. I may take myself to be perceiving an expanse of red, but it does not follow that I am, for I might be mistaken about what the colour is. I may be the victim of some strange neurosis such that whenever I see purple I think it is red, and this may even extend to my thinking I hear people talking about red when in fact they are talking about purple. It is even possible that everyone should make such a mistake about some particular expanse of red, because, let us suppose, there is some strange gas in the air, or some persistent hypnotist in the vicinity. It never follows logically from what people take themselves to perceive that they are, in fact, perceiving such things. It is at this point that the reduction of knowledge of the external world to knowledge derived from the perception of the particular moment breaks down. Nevertheless I think there is some truth in this attempted epistemological construction, for there is another, weaker, type of construction that we have yet to consider.

## 12.3 DERIVATIONS

Reductionism/constructivism is the attempt to show how we know what we do by deducing, logically, one kind of knowledge from another. The attempt to deduce knowledge of reality from knowledge of perception-that failed, but perhaps it failed because we were asking too much. We have said that the epistemologist is primarily interested in the right to be sure, and we have seen (10.5) that to show that we have the right to be sure of something is, *ipso facto*, to show that we have the right to be sure that it is true, and hence to provide an acceptable, although not a logically conclusive, proof that we know. So perhaps we should consider not, as with reductionism/constructivism, whether what we know at one level logically guarantees our knowledge (i.e. both the truth and our right to be sure of what we claim to know) at another level, but whether it logically guarantees our right to be sure. What we would then be trying to deduce, logically, from knowledge at a lower level would be not the truth of what is known, but our right to be sure of it—this would be a construction, or conversely a reduction, of the right to be sure rather than of knowledge as such. I will call this a 'derivation' of the knowledge: knowledge $X$ is derived from knowledge $Y$ if knowledge of $Y$ is sufficient to give us the right to be sure of $X$, just as knowledge is derived from perception if perception is sufficient to give us the right to be sure of whatever it is.

An example may help. The knowledge that the *Grand Canyon Daily Echo* is a reliable paper, together with the knowledge that today's issue reports that Martians have landed in Manchester, does not entail the knowledge that Martians have, in actual fact, landed in Manchester. Even reliable newspapers can be, and often are, wrong. Nevertheless the fact that I know that the *Grand Canyon Daily Echo* is a reliable paper, together with the fact that I know that today's issue reports the landing of Martians in Manchester, does entail that I have the right to be sure—whether or not I actually am sure—that the Martians have landed (given, I suppose we should add, that the reliability of the paper is sufficient to outweigh the implausibility of its report). The difference is that I can have the right to be sure of $p$ even where $p$ is false, whereas I cannot know $p$ if $p$ is false. The old hope that Cartesian scepticism might be refuted by providing a logical reduction of all knowledge to some basic type of absolutely certain knowledge is due, in part, to the failure to see that the right to be sure can be guaranteed in circumstances where truth is not, that what is logically guaranteed is the right to be sure rather than knowledge (which involves truth as well as the right to be sure). We have seen that the same mistake leads to one form of scepticism, to the lament that since what gives us the right to be sure does not logically guarantee the truth of what is known it cannot give us knowledge after all.

Once we focus our attention on the right to be sure, rather than on the truth, of what is known it may seem that we can easily show how sense datum knowledge can be derived from knowledge of perception-that, and knowledge of reality from sense datum knowledge. What gives me the right to be sure that I am perceiving a certain sense datum, e.g. of a book? Surely it is the fact that I take myself to be perceiving a book and so, implicitly, take myself to be perceiving a book-like sense datum, that gives me the right to be sure that I am perceiving such a sense datum. Of course the fact that I take myself to perceive a book doesn't mean that I do perceive one, for I may be misled by some strange play of light on the table in front of me, but the chances of this are so slight as, apparently, not to affect my right to be sure. And similarly what gives me the right to be sure that such a book really exists is, surely, the fact that I perceive the book—or if you prefer, the fact that I perceive a book-like sense datum. Once again the fact that I do perceive a book-like sense datum doesn't mean that there is a book there, for it might be a hallucination, but the chances of this are so slight as, apparently, not to affect my right to be sure. Thus my knowledge of perception-that, that I take myself to perceive a book, seems to entail first my right to be sure that I am perceiving

OUR KNOWLEDGE OF EXTERNAL EXISTENCE 217

the appropriate sense datum, and through it my right to be sure that such a book does really exist.

This is all very swift and simple, but unfortunately too swift and simple. No one would deny that if one takes oneself to perceive a book one has the right to be sure that one does, in fact, perceive a book, and so in turn the right to be sure that there is a book there. The question is whether these implications are entailments. I do not think they are, for it is possible to imagine circumstances in which we would not allow that these things did give us the right to be sure. Suppose that our senses were much more deceptive than they are, that mistakes about what we perceive were the rule rather than the exception—would we then say that taking oneself to perceive something gave one the right to be sure that one was perceiving it, or in turn that it really existed? Obviously not; it is only because our perception is normally veridical, because ordinarily we do not sense what is not really there, and do not make mistakes of perception-that, that we allow that perception and perception-that give us this right to be sure. My right to be sure that this book really exists follows from the fact that I know I perceive a book-like sense datum, only given that perception is normally veridical. So if we are to derive my knowledge that the book really exists from the sense datum knowledge, we will have to show how sense datum knowledge entails the right to be sure that perception is normally veridical. Similarly if we are to derive the right to be sure that I perceive a book-like sense datum from knowledge of perception-that we will have to show how knowledge of perception-that entails the right to be sure that normally we do perceive what, including the sense data, we take ourselves to. Let us begin with the latter.

12.4 A DERIVATION OF SENSE DATUM KNOWLEDGE

Perception-that ordinarily goes beyond what is immediately perceived, usually (as when I take what I perceive to be a book) including a claim that what is perceived does really exist and possesses various qualities and features which I do not, at the time in question, perceive. But such a claim can be said to involve an implicit judgement about what is immediately perceived, it being what is immediately perceived, the book-like sense datum, which is, in an obvious sense, my basis for the claim that there really is a book here. Now at the moment we are concerned with the perception-that restricted to what is immediately perceived, the implicit judgement that I perceive a book-like sense datum. And our question is: what gives me the right to be sure that I do perceive the sense datum I implicitly take myself to?

The answer may seem obvious: the fact that I perceive it.[1] But this reply ignores the crucial point at issue: what gives me the right to be sure that I do perceive it, and do not merely 'perceive' it, take myself to perceive it when I do not? Again, it is tempting to say that what gives me the right to be sure that I am perceiving a book, and hence that I am immediately perceiving the appropriate sense datum, is that I know what books, colours, shapes, etc., are like. But what gives me the right to be sure that books, colours, shapes, etc., are like this? Presumably the fact that I have perceived them so often. And what gives me the right to be sure that I have perceived them so often, and not merely 'perceived' them? We are back where we began.

If we are ever to make the step from what we take ourselves to perceive to what we actually do perceive, we will need a case where perception of something is, by itself, sufficient to give us the right to be sure that we perceive, and not merely 'perceive', whatever it is. This step can, I think, be made by use of the notion of 'immediate appearance' (11.5). The fact that I perceive what I do does, by itself, give me the right to be sure of the immediate appearance of what I perceive, that what I perceive is, as I perceive it to be, 'like this'. Yet to say that I have the right to be sure of this is to say very little. It is simply to say that what I perceive is as it is, and since this is to say virtually nothing I can hardly say that I know it, for to know this is to know virtually nothing. If my knowledge is to have any content, if I am genuinely to know anything, then I will have to classify what I perceive, say not simply that it is as it is, but that it is of some particular sort. And, it seems, the moment I attempt to classify what I perceive the same old question arises: what gives me the right to be sure that this classification is correct?

Consider what is involved in classifying something as, say, red. Two separable things can be involved here, although in adult, language-using, human beings they normally go together. I may be saying that what I perceive is of a certain sort without saying or

[1] 'What better answer could he give than that these just were the experiences he was having?' Ayer VI, p. 69.

OUR KNOWLEDGE OF EXTERNAL EXISTENCE 219

knowing what things of that sort are called; a man who has never learnt a name for kangaroos may say 'Here's another of those strange things—I wish I knew what they are called', but in classifying it along with those other strange things he is, in a definite sense, classifying it as a kangaroo. Or I may be saying that what I perceive is of the sort called $x$ without saying or knowing what $x$s are; the lab cleaner who has been told that this is an oscillograph may say 'Here's another oscillograph—I wonder what oscillographs are', but in calling it an oscillograph he is, in a definite sense, classifying it as an oscillograph. We might distinguish these two cases by the use of quotation marks, distinguishing classifying something as an $x$ or knowing that it is an $x$ from classifying it as (called) '$x$' or knowing that it is (called) '$x$'. Two points need to be noticed: first, to repeat, these two things normally go together, and second, in saying that a person knows that this is 'red', i.e. of the sort called 'red', we do not necessarily mean that he knows this English word—any synonym for the quoted word will do, for what is necessary is that he knows some name for the type of thing in question.

However we are not here concerned with the learning of language, so it is only with the classification of something as $x$ and the knowledge that it is $x$, as opposed to the classification of it as an '$x$' and the knowledge that it is an '$x$', that we are concerned. Our question is not 'What gives me the right to be sure that what I perceive is "red"?' but 'What gives me the right to be sure that what I perceive is red?' Now to classify something as red is to say that it is, in a certain respect, like various other things, e.g. roses, tomatoes, the front door. This is not, as we have seen (11.5), to say that 'This is red' *means* 'This is, in the relevant respect, like roses, tomatoes and the front door', because the truth of the former is compatible with roses, tomatoes and the front door being indigo, even with their not existing at all. What it means is that this is, in the relevant respect, like such things as are, as it happens, red, and, as it happens, roses, tomatoes and the front door are red. So in order to know that what I perceive is, as I take it to be, red, I have to know that what I perceive is, as I take it to be, like other things which happen to be red. In talking about the immediate appearance of what I perceive I do not classify it in this way; I say merely that what I perceive is 'like this'. However I have perceived many things that look red, and in each case I have the right to be sure that they are 'like this', where the 'like this' in fact refers to their redness. So I have the right to be sure that I have perceived *a*, *b* and *c* where *a*, *b* and *c* look, as it happens, red. Moreover redness is an appearance-determined quality such that all I have to do to discover what colour a colour is is to see it. So I have the right to be sure that *a*, *b* and *c* are all alike in this respect, in their 'like thisness'. And what more is necessary for my having the right to be sure that this, which I now see, is red, as opposed to having the right to be sure that it is 'red', than that I have the right to be sure that this is, in the relevant respect, like *a* and *b*, i.e. that this is of the same kind (known in English as 'red') as *a* and *b*?

In this way I have the right to be sure that what I perceive is in appearance as I take it to be, in so far as I have previously perceived other things which are like it in appearance, and in so far as I am taking what I now perceive to be like those things in appearance: I perceive something. I cannot be mistaken about its immediate appearance. I perceive something else. I cannot be mistaken about its immediate appearance. I take the immediate appearance of what I now perceive to be, in a certain respect, like the immediate appearance of what I first perceived. This is, to some extent, to classify what I now perceive, and in so far as it is a classification it may be mistaken. I may be wrong in taking what I now perceive to be like what I first perceived; my memory may be playing tricks on me. But the fact that I may be wrong does not mean that I cannot have the right to be sure, and here, for once, we have a case where my perception does give me the right to be sure that I perceive whatever it is and do not merely 'perceive' it, because redness is an appearance-determined quality which means that perception by itself can tell us that two instances are alike. So it follows logically from the fact that I take myself to be perceiving something whose immediate appearance is classifiable along with the immediate appearance of something else that I have the right to be sure that I am, in fact, perceiving something of that sort. It does not follow, logically, that I *am* perceiving something of that sort, for what I perceive may not, after all, be like what I previously perceived. But I do have *the right to be sure*, because my previous perception gave me the right to be sure that what I perceived was 'like this' in immediate appearance, and my present perception gives me the right to be sure that what I now perceive is 'like this' in immediate appearance. And with appearance-

## OUR KNOWLEDGE OF EXTERNAL EXISTENCE

determined qualities all that is necessary for the right to be sure that two instances are alike is that we perceive them.

Moreover a sense datum can be said to consist of appearance-determined qualities. The sense objects themselves are appearance-determined and in ascribing other qualities to the sense datum we are, as we saw (11.7), restricted to the apparent qualities of what we perceive, i.e. the qualities that what we perceive appears to have, and these, obviously enough, are appearance-determined also. So the present argument entitles us to say that a person necessarily has the right to be sure that he is perceiving the sense datum he takes himself to perceive, so long as this taking himself to perceive a certain sense datum consists solely in his taking himself to perceive a sense datum which is, in various respects, like other sense data he has taken himself to perceive. From here, perhaps, we should go on to consider how we have the right to be sure of sense datum statements which go beyond this, which classify what is immediately perceived in terms of what the perceiver has not immediately perceived. But this would take us too far from the main enquiry; let us be content with the more modest point.

I think it is in this way that we can provide an answer to a regressive scepticism. What we come down to, in the end, is what we take ourselves to perceive, and the sceptic cannot sensibly ask what gives us the right to be sure of that, for it is not the sort of thing we can be mistaken about. We answer the sceptic by showing that we have the right to be sure, that our right to be sure follows from something about which we cannot be mistaken. Even so a loophole remains. The fact that I take myself to perceive something which is 'like this' in immediate appearance is sufficient to give me the right to be sure that I do perceive something 'like this' in immediate appearance, but the fact that I once took myself to perceive something which was 'like this' in immediate appearance does not, by itself, give me the right to be sure here and now that I did perceive something 'like this' in immediate appearance. I have to *remember* that I took myself to perceive it and what, the sceptic will ask, gives me the right to be sure of that? This is a good question, but it is one that is irrelevant to our present enquiry. We are concerned with perception, not memory, and we have shown how perception provides the right to be sure. The question of how we remember what we perceived, of what gives us the right to be sure that we did perceive, or at least took ourselves to perceive, what we remember ourselves perceiving or taking ourselves to perceive, belongs to another part of epistemology.

## 12.5 A DERIVATION OF KNOWLEDGE OF REALITY

Our next question was: what gives me the right to be sure that the things I perceive do really exist, or, as we are putting it, that the things I immediately perceive include parts or aspects of external objects, that the sense data I perceive are sense data of real existents? We have decided to accept the common sense view that we do perceive external objects; the question is how do we, as we undeniably do, know that we do, what gives us the right to be sure that this is so? One way of answering the question would be to consider what would have to be different about what we perceive for us to say that we no longer have the right to be sure that we perceive external objects. And one way of showing that the question is neither idle nor absurd would be to point out that if what we perceived were different in various ways we would no longer allow that we did have the right to be sure.

The feature of what we perceive which leads us to interpret our perception as perception of an external world is its constancy and coherence, to use Hume's terms, its unity, to use Kant's. Our perception is not perception of an abruptly altering succession of unrelated and random flashes, bangs, lines, shapes and colours. Instead it has a definite continuity, an order and cohesion such that what we perceive via a particular sense at a particular time fits not only with what we perceive via other senses at the same time, but also with what we perceive via that, and other, senses at other times. It is this which leads us naturally, perhaps inevitably, to interpret our perception as perception of determinate, though alterable, persisting, spatio-temporally continuous objects; and it is this which leads us naturally, perhaps inevitably, to interpret our perception as perception of such objects from a determinate, though alterable, point of view. It is this constancy and coherence, the fact that our perception fits and relates together from one moment to another and from one sense to another and, as we tell from our perception, from one person to another, that we accept as giving us our right to be sure that our perception is perception of an external world. It does not, of course, give us the right to be sure that our perception is always perception of external objects, although no doubt it would do this were it not for the fact that we all well know that perception is not always veridical.

But if, on the other hand, our experience were not like this, if this constancy and coherence in what we perceive were missing—perhaps not entirely but to a large extent—we would not come to think of our perception as perception of external objects, and we would not have the right to be sure, as we do, that we do perceive an external world. The question of which changes in this unity of what we perceive would have which results on our conception and our right to be sure of the external world is a difficult and fascinating one, but one which is too big to tackle here. Perhaps Kant has been the only one to attempt, if indirectly, a thorough-going answer.[1]

Instead of answering it I want to say a little more about the question itself. It will not do simply to ask, as Kant did, 'What makes experience of an external world possible?' or, as Strawson does, 'What are the conditions for a non-solipsistic conceptual scheme?' for these questions are ambiguous. We might be asking 'What are the features in what we perceive which explain our interpreting it as an external world, in terms of a non-solipsistic conceptual scheme?'; or we might be asking 'What conditions have to be satisfied before what we perceive can, logically, be interpreted as an external world?' The difference is that the first question is concerned with the conditions that make it empirically possible for us mortal men to interpret what we perceive in this way, whereas the second question asks what the logically necessary conditions are. Now whether or not it is, in fact, possible for us to interpret what we perceive in a certain way seems to depend as much on us, our intelligence, practical needs, interests, etc., as on what we perceive. It may also depend on what knowledge and concepts we possess already, for it may be possible for *us*, drawing on the conceptual scheme we have developed for application to what we now perceive, to interpret some new and unusual perceptual world, e.g. Strawson's auditory universe,[2] in terms of a non-solipsistic conceptual scheme, when a being with no previously acquired concepts and knowledge would be extremely unlikely to do so. I have argued elsewhere[3] that in his discussion of this audi-

---

[1] The best recent discussion seems to be Price II; cf. also Ayer II, § 15; Strawson I.
[2] I, ch. 2.
[3] I.

tory universe Strawson fails to separate such questions as 'Can we develop a non-solipsistic conceptual scheme which will apply to the auditory universe?', 'Could the inhabitant of such a universe develop and apply such a non-solipsistic conceptual scheme?', and 'Would such a non-solipsistic conceptual scheme be of any use in such a universe?'. And this, I think, is due to a failure to distinguish the question of what, as a matter of fact, brings us to interpret what we perceive as we do, from the question of what conditions have to be satisfied before such an interpretation is logically possible.

Kant, I would suggest, made a similar mistake. He argued that the necessary condition for our perception's counting as perception of an external world was that what we perceive should exhibit a certain unity, and he tried to set out what the essential features of that unity are. Now it is undeniable that there are certain features of what we perceive which enable us to interpret it in this way as an external world, but Kant overestimated not only the extent to which these features, such as causal connection, are necessary for such an interpretation, but also the extent to which these features, and so the unity itself, are responsible for the interpretation. Kant talks as though if what we perceive exhibits these features our perception must be construed as perception of an external world, when the point seems rather to be that although these features lead us to interpret what we perceive in this way it remains possible to interpret it in some other way, e.g. that of the Idealist. Nevertheless Kant is aware, if confusedly, of the fact that our conception of what we perceive as an external world depends in large part on how we interpret what we perceive, and is not just a necessary consequence of the way what we perceive is perceived to be. For he thinks that the features which lead us to interpret what we perceive as we do are, somehow, added by the perceiving mind to what is perceived. So instead of saying that it is the unity in what we perceive that leads us to interpret what we perceive as an external world, he seems to be saying that it is the interpretation which produces the unity. This seems to be due to a failure to distinguish the interpretation from the possession by what is perceived of those features which make the interpretation possible and natural.

However the main point is the distinction between the conditions which are logically necessary for our interpreting what we perceive as an external world, and the conditions which are empirically necessary if we mere humans are so to interpret what we perceive. The question of the empirical conditions seems not to be a philosophical one, although its answer has, no doubt, philosophical consequences. It is, presumably, for psychologists, anthropologists, sociologists and the like to tell us why we interpret what we perceive as we do, what the point of our so interpreting it is, and so on, and so tell us what a being would have to be like before it would interpret what we perceive in the same way. But, on the other hand, if, as philosophers, we restrict ourselves to the question of what conditions are logically necessary we seem to miss the point of questions like those of Kant and Strawson. Indeed I have argued[1] that we could, logically, interpret our perception as perception of an external world almost, but not quite, no matter what what we perceive is like. All that seems necessary is that we give some currency to the notion of places outside our present range of observation, and of things existing at those places. And it is possible for us, by borrowing heavily from the conceptual scheme established with reference to what we now perceive, to apply, admittedly arbitrarily, these notions to even the most chaotic and disorganized experience. The interesting question of how what we perceive provides for a non-solipsistic conceptual scheme seems to have vanished somewhere between the empirical question of why and how a particular being would develop such a scheme, and the purely conceptual question of what concepts have to have application is such a scheme is to be used. It seems to me that the question to ask here is not 'What makes this interpretation of what we perceive *possible*?', but 'What makes this interpretation *justifiable*?' Or, more illuminatingly, 'What does what we perceive have to be like before we would, given those standards of having the right to be sure which are encapsulated in our accepted use of the verb 'know', have the right to be sure that this interpretation of what we perceive is the correct one?' This question avoids both the irrelevant reference to the intelligence and needs of the being whose perception it is, and the irrelevant reference to the formal conditions which have to be satisfied before such an interpretation can be made. It seems to me to be the important question which Kant and Strawson are raising. Similarly it seems to be the question Hume was asking when he asked 'What causes induce us to believe in the existence of body?'. In so far as this is a philosophical

[1] I, pp. 527 ff.

epistemological, question, and not a psychological question, it is: 'What gives us the right to be sure that we perceive external objects?'

However, let us return to the main argument. We have said that it is the constancy and coherence in what we perceive, the unity of our experience, which justifies our interpreting our perception as perception of an external world, i.e. our adoption and use of a non-solipsistic conceptual scheme, and so our right to be sure that we perceive external objects. We don't want to say that we have the right to be sure that we *always* perceive external objects, much less that our perception is always veridical, for we are well aware that this is not the case. Rather what we have vaguely, and without further analysis, referred to as the constancy and coherence of what we perceive provides us with a general, though rebuttable, presumption that what we perceive on some particular occasion really exists. Just as it would be wrong to say that we have the right to be sure that all our perception is perception of external objects, so equally it would be wrong to say that we do not have the right to be sure that what we perceive on some particular occasion really exists until we have, so to speak, tested it for constancy and coherence. Ordinarily we no more question the real existence of what we perceive than we ask people who are introduced to us to show us their birth-certificates. Perception by itself, so long as there is nothing strange or unusual about it, and no factors affecting it that we know of, is normally regarded as sufficient evidence for the real existence of what is perceived. Such things as the vagueness or strangeness of what is perceived ('Snakes on the counterpane? I must be seeing things'), or the evidence of our other senses, or our knowledge of some special psychological or physiological state that we are in ('I thought I saw a lion but I am so jittery I wouldn't be sure') can take away this right to be sure, but so long as everything seems straight-forward, perceiving something does, by itself, give us the right to be sure that that thing does really exist. Yet although we have this general, though rebuttable, right to be sure that what is perceived really exists, this is not to say that we have a general, even rebuttable, right to be sure that our perception is veridical. For we have seen (6.1) that veridical perception also involves making no mistakes about what we perceive, no mistakes of perception-that. And we do not have a general, even rebuttable, right to be sure that things are as we take them to be. Mistakes of this sort are far more common than perceiving something that doesn't really exist, and one does not have the right to be sure that what one perceives is as one takes it to be unless one is familiar with things of this sort or, perhaps, this particular thing.

This leads to the question of how, in times of doubt, we determine whether what is perceived does or does not really exist. The answer is familiar. We decide by discovering whether what I now perceive fits in with what I perceive by other senses and at other times, and with what other people perceive, or, if you prefer, whether I perceive other people talking and acting as if they too perceive what I do. The fact that all these tests work by referring to what is perceived by me and others makes it tempting to say that the question of whether what is perceived really exists or not is equivalent to the question of how present perception relates with other perception, and so leads us to a Phenomenalistic analysis of real existence. Certainly the Phenomenalist is correct to the extent that it is by comparing what I perceive on one occasion with what I and others perceive on other occasions, that I resolve any doubts I may have about the veracity of my senses (if the question arises in a situation where I cannot carry out such tests—did something flash by me just now or did I imagine it?—or if I refuse to be convinced by the tests I do carry out, I just have to remain undecided). Nevertheless we have seen that statements of real existence are not logically equivalent to statements about what is or can be perceived, unless the latter presuppose the notion of real existence.

It has been suggested[1] that the connection between accounts of what is perceived which involve no reference to real existence, and accounts of what really exist, is like that between evidence and verdict. This is a helpful analogy. We can say that perception of $x$ is excellent *prima facie* evidence for the existence of $x$, but that further evidence may lead us to reverse the verdict. And once a certain amount of evidence is available no further information could lead us to reverse the verdict, unless that further information also affects the accuracy of the evidence (e.g. by showing that I did not, after all, see others talking and acting as if they too saw it). Austin[2] has objected that this account distorts what goes on when we come to think that something really exists, but the question is

[1] Warnock I, ch. 9.      [2] II, pp. 141–2.

not how, normally, we decide that what we perceive really exists (answer: normally we don't decide, we just, and justifiably, take it for granted that it does), but how, theoretically, the assertion that what we perceive really exists can be justified. I think a more important objection is that Warnock exaggerates the logical gap between accounts of what we perceive, the evidence, and existential statements, the verdict. He thinks that the evidence can never entail, although it can justify and explain, the verdict, but we have seen that, for example, the fact that so many people have in so many ways and at so many times perceived the island Great Britain does entail that this island does exist. Warnock is misled here by his mistake about statements which describe what we perceive without saying anything about real existence; his 'It seems to me as if . . .' statements describe not what is perceived, but what the perceiver takes himself to perceive (cf. p. 176 n and p. 195 n above).

## 12.6 THE EPISTEMOLOGICAL MORAL

We have been indulging in that old-fashioned, some would say disreputable, business of constructing the external world from our sense data. Our aim was to show that Empiricism might well be true by showing how perception, by itself, could give us the right to be sure of the existence of external objects. I think our answer also does something to show how our perception is responsible for our conception of the external world. In showing what it is in what we perceive that is accepted as providing us with the right to be sure of the external existence of what we perceive, we also show what it is in what we perceive that is responsible for our thinking of what we perceive as an external world.

We saw that the attempt at a logical construction of our knowledge of the external world fails. Instead we concentrated on what I called a 'derivation' of this knowledge, on showing how what we perceive guarantees, logically, our right to be sure of the real existence of what we perceive, even though it does not logically guarantee that it does really exist. By beginning with immediate appearance and what a man takes himself to perceive we showed how we have the right to be sure that we immediately perceive various things, and we then saw how the constancy and coherence of what we immediately perceive gives us the right to be sure that the things perceived do really exist. Now in part our answer to the question 'How do we know that external objects exist?' has been 'Because our concept of Knowledge is as it is'. This may seem unsatisfactory, pointless, trivial, even circular, but in a way it is the only answer we can expect. If it is true that we know certain things it must be because the various conditions which have to be satisfied before we can speak of knowledge are satisfied, and to explain what these conditions are is, in its way, to explain the concept of Knowledge. If it is true that we know that external objects exist it must be that our concept of Knowledge is such that our 'evidence' (whatever it is that leads us to say we know) for the existence of external objects is sufficient to satisfy the conditions for knowledge implicit in that concept. The interesting and epistemologically important question is the question of what it is that thus satisfies these conditions—this is the epistemological, as opposed to the psychological, 'How do we know . . .?' We must not be disappointed when the answer to the further question 'How is it that these things do satisfy the conditions for knowledge?' is merely 'Because our concept of Knowledge is as it is', for that is the only answer possible. The moral is that epistemology is primarily descriptive, concerned with discovering, with reference to the accepted concept of Knowledge and so the accepted conditions for having the right to be sure, what it is that is accepted as providing us with knowledge, giving us the right to be sure. It would be a mistake to think that epistemology can add to or subtract from our existing body of knowledge. Rather it is concerned with discovering how the things that do belong to this body qualify as belonging.

I think there is another moral to be drawn from the fact that epistemology is concerned with our right to be sure of what we claim to know rather than with its truth. It is often said that 'How do we know?' comes closer to the epistemological point than 'Do we know?' We can now see that this latter question is answered not by proving beyond all possibility of a doubt that what we claim to know is true, but by trying to prove, beyond all possibility of a doubt, if you like, that we have the right to be sure that it is true. This involves discovering what it is, in the particular connection, that is accepted as providing the right to be sure, and so enables us to throw light both about the nature of the knowledge and on the concept of Knowledge itself. Traditional epistemological topics of induction, necessary truth, laws of nature, the past, other minds, and so on are more amenable to treatment if we try to show not that what we claim to know is true beyond all doubt but that we have, beyond all doubt, the right to be sure of its truth.

## 12.7 OUR KNOWLEDGE AS AGENTS

There is one radical objection to these constructions from sense data and, indeed, the entire Empiricist theory. It is that Empiricism is a 'paralytic's epistemology', induced, no doubt, by an overdose of ghost in the machine metaphysics. Traditional Empiricism, it can be argued, tended to follow Descartes in identifying the person with a mind or consciousness or thinking substance which in some more or less mysterious way learns about the world via the sensory mechanism of a particular body. Thus we are thought of as passive receptors of information about an outside (hence 'external') world, although it is allowed that we can actively correlate, arrange and form inferences from this information. But, the reply is, this is to forget that human beings are, among other things, physical bodies which can and do act and move, and it is from what we do as much as from what we perceive that we learn about the world in which we find ourselves, and come to think of it as we do. It isn't just because I perceive a table but, perhaps more importantly, because it gets in my way, that I say that a table exists. If I am in doubt about its existence I don't call in the evidence of other senses or other perceivers as the Phenomenalist suggests, I simply follow Dr Johnson and kick at it, and if I bang my toe then its real, external, existence has been established. It's because I can and do move from place to place that I learn of the existence of places where, at the particular moment, I am not, and so of the existence of objects at those places, objects which at the particular moment I do not perceive. It's because I can and do move round the table and notice changes in its appearance as I move, that I learn that there is more to it than what I perceive of it on any particular occasion. And if I'm in doubt whether this telephone really exists, better than touching, tasting, smelling it I might just pick up the receiver and try to dial TIM.

It is undeniable that philosophers have often distorted the facts by overemphasizing the contemplative aspects (thinking, reasoning, perceiving) of the human mind and underestimating the fact that man is an active agent. Indeed it is well nigh impossible to exaggerate the importance of our movements and actions in our coming to know and conceive of the world in the way that we do. But, as always, we are concerned with the theoretical consequences, in this case the consequences that the fact that we move and act has for any account of the source and scope of our right to be sure of the existence of external objects. Our question is whether this fact that we move and act provides us, in any special way, with our right to be sure that there exists an external world. And at first sight this seems doubtful.

For the fact that something blocks our motion does not mean, logically, that there is real existent before us, as the example in 6.2 shows. However such hallucinations do not occur, so it is true, as a matter of empirical fact, that whenever our motion is blocked there really exists some thing which blocks it. Thus given that we have the right to be sure of this generalization, the fact that our motion is blocked logically guarantees our right to be sure that there is some real existent blocking it. But as the Phenomenalist will cheerfully point out, our right to be sure of this empirical generalization itself comes from what we perceive when our motion is blocked, including, most importantly, the fact that we perceive other people talking and acting as if their motion, too, is blocked in the same place. In fact how do I know that my motion is blocked in the first place? Surely it is from what I feel, from the fact that when I try to move in a certain way I feel something resisting me, together, perhaps with the fact that I see a part of my body held firm against the object in question. Similarly, how do I know that I do in fact lift the receiver and dial TIM? Surely from what I perceive.

In other words although the fact that we move and act is important in explaining how we acquire individual pieces of knowledge, it makes no difference to our epistemological account, since although these movements and actions may be responsible for our perceiving what we do, it is still our perception that provides us with our knowledge, is the source of the right to be sure. If the fact that we are agents is to be of any importance for the theory of knowledge, as opposed to the practice of gaining it, it must be because, as agents, we have knowledge and acquire the right to be sure in a way that is independent of the fact that we are also perceivers, i.e. in a way that does not rely on our perception of where we are and what we are doing.

It has been suggested that this is in fact the case, that there is a special type of knowledge not derived from perception, viz. 'non-observational knowledge' such as we have of the position and movement of our limbs, and of our intentional actions.[1] I have argued elsewhere[2] that there is a difference between our knowledge of what we intend to do or be doing and our knowledge that we actually are doing what we intend to do (both of which might, with confusing results, be called 'knowledge of our intentional actions'), and that a failure to distinguish the two leads to the notion of 'non-observational knowledge' covering two different types of knowledge, which I call 'introspective knowledge' and 'subliminal knowledge'.

My knowledge of what I intend to do is introspective in a traditional sense (although I wouldn't want to commit myself to any traditional doctrine of introspection). I know, without having to perceive or examine anything, what I intend to do or be doing. Now this 'introspective' knowledge of what I intend to do does not by itself constitute knowledge or give me the right to be sure of what I am doing. It does not follow from the fact that I intend to be doing $x$ that I am doing it, nor even that I have the right to be sure that I am doing it. I may, for example, believe it possible to throw a ball a thousand miles, so I may intend to be doing it, and know that I intend to be doing it, but I do not have the right to be sure that I am doing it. I have the right to be sure of it only if I have the right to be sure that I can do it when I try, and although I may be fool enough to think I can do it if I try, I certainly don't have the right to be sure that I can. In order to have the right to be sure that I can do something I have to have the right to be sure that I have succeeded in doing it or similar or more difficult things in the past, and in order to have the right to be sure of that I have to have noticed or discovered what has happened in the past when I tried to do such similar or more difficult things. More is needed than mere knowledge of intentions, and this is true even where the action is some slight bodily movement, such as waggling a finger or twitching an eyebrow. It is only from experience that we know whether we can move the various parts of our body, whether we can waggle our fingers or our ears, twitch our eyebrows or our kidneys. 'Introspective' knowledge does not provide any non-perceptual way of knowing about what is going on in the external world.

[1] Cf. Anscombe I.    [2] III.

'Subliminal knowledge' is knowledge based on the unconscious or subliminal registering, as opposed to the conscious noticing or perception, of various sensory cues. For example I may know that my hand is moving not because I see it or even feel it, but because of various tactual and kinaesthetic sensations which I do not consciously notice, and hence do not perceive, in any full sense, at all. Knowledge of the position and movement of our limbs is often of this sort, although it is not necessarily so, for the sensory cues which provide the knowledge can, if necessary (and it sometimes is), be consciously noticed and so perceived. Now this subliminal knowledge is knowledge of the external world which is not sense-knowledge. For sense-knowledge, as we defined it (10.6), is knowledge for which perception of some specific thing or things is necessary, and I can, for example, know that my hand is moving without consciously perceiving it, without noticing or perceiving the kinaesthetic sensations which, subliminally, provide me with this knowledge. Notice, however, that this subliminal knowledge, by itself and independent of any sense-knowledge, is very meagre indeed. It does not extend to such things as my knowledge that I am lifting a telephone receiver and dialling TIM. Obviously I do not have the right to be sure of that unless I see, or perhaps feel (paying careful attention to where I put my fingers, what it feels like, etc.), what I am doing. Even with a simple movement like my arm's moving up and down, where this is something done to me rather than something I do, it seems that I only have the right to be sure if I consciously notice the kinaesthetic and tactual sensations involved. Subliminal perception, by itself, provides us with only the roughest knowledge of where our limbs are and what they are doing. Of course one can, with intentional movements, know precisely what one is doing, without seeing or consciously feeling what is going on. But this is only because they are intentional movements and one has the prior introspective knowledge that one intends to be performing these movements, and the prior sense-knowledge that these movements are within one's capabilities. Wittgenstein says[1] 'I let my index finger make an easy pendulum movement of small amplitude. I either hardly feel it, or don't feel it at all. Perhaps a little in tip of the finger, as a slight tension. (Not at all in the joint.) And this sensation advises me of the movement?—for I can describe the movement exactly'.

[1] I, p. 185.

Certainly the slight sensation does not, and could not, give me my precise knowledge of what I am doing. It is my knowledge of what I intend to be doing, together with my knowledge that this is a movement well within my capacities, which gives me such precise knowledge. The kinaesthetic sensation functions only as a sign that everything is going to plan. In and by itself subliminal knowledge is very scanty and very vague.

Nevertheless it does occur, and it is knowledge of the external world which is not sense-knowledge. It follows that the claim that all knowledge of the external world is sense-knowledge is, interpreted as we have interpreted it, incorrect. If we cared to define sense-knowledge not as knowledge where perception of some particular things is a necessary condition of our acquiring the right to be sure in the way that we do, but as knowledge where sensory awareness, conscious or not, via some sense-modality, of some particular things is a necessary condition of our acquiring the right to be sure in the way that we do, this subliminal knowledge would then count as sense-knowledge. Apparently the existence of this type of knowledge calls not so much for the rejection as for the modification of Empiricism, as we have described it. We can make this modification if we like, but we should also notice that this subliminal knowledge affects not only our definition of Empiricism but also the Empiricist description of our knowledge in terms of a hierarchy of types of sense-knowledge (12.1). For we have here a case where knowledge of reality (e.g. that there is a hand—mine—in a certain place), is not based on sense datum knowledge; nor on knowledge of perception-that.

Is there any other subliminal knowledge of the external world, besides this knowledge of the position and movement of our limbs? I do not think so. We might consider such things as the so-called sixth sense, as when I realize that someone has entered the room not because I have consciously perceived anything that tells me so but, presumably, because of the unconscious registering of some slight movement, sound or smell, or when I realize that this is the tenth chime of the clock, even though this is the first that I have consciously heard. Such things undoubtedly happen, but the question is whether we should speak of knowledge. Would we say that I have the right to be sure, and so know, that someone has entered the room, that this is the tenth chime? If we would I think we would also be inclined to add that this is a loose, extended, even inaccurate, use of the word 'know', rather like that in which we might say that the fortune teller knew she was going to marry a tall, dark man, just because what the fortune teller said turned out to be correct. In cases like this we are usually chary about allowing the right to be sure. The fact that we are correct in thinking something to be the case encourages us to say, with hindsight, that we did have the right to be sure, but the strangeness and unfamiliarity of the method of arriving at the truth inclines us to say that we did not know. If these phenomena—the sixth sense or accurate fortune telling—were much more common than they are, as common, say, as that of a man's knowing what he is doing without looking to see, then I think we would allow that we had the right to be sure. But these phenomena are not common enough to be reliable, and so I do not think that we would say that we had the right to be sure. It seems, then, that although subliminal knowledge does not in fact extend beyond knowledge of the position and movement of our limbs we can imagine circumstances where we would allow that it included knowledge of other things.

Thus the fact that we can move and act does make some difference to our account of our knowledge of the external world, in particular our knowledge of that external object which is our own body, although it would be going a little too far to speak of action as a possible source of knowledge as a rival to perception. But although the fact of our motion makes but slight difference to our account of knowledge it makes an enormous difference to any account of how we come to conceive of the world in the way that we do. The very notion of 'external' existence, of something outside me, it can be said, is due to the fact that we come into contact with things which restrain and restrict our actions and movements. Indeed the moment we are prepared to talk of ourselves as moving about and coming into contact with things, that moment we are forced to reject an Idealist interpretation of what we perceive, and thus the Idealist theory of perception. According to the Idealist what we describe as our motion can be nothing more than a change among entities which exist only in our consciousness. In so far as we want to say that we move among things, that they move around us, that we bump into them and they into us, that we notice changes in their appearance due to our or their movement, so far we are forced to reject an Idealist theory. This is another point at which the Idealist parts company with common sense, and, I think, it counts more against the Idealist than his denial of the real, non-sense-dependent existence of what we perceive. For although we all want to say that the table continues to exist when we go out of the room, we are made slightly uncomfortable about saying this when we consider the sort of argument that Berkeley raised against Locke. We all want to say that it still exists, but surely some crucial evidence is, necessarily and *ex hypothesi*, missing? Isn't the whole thing something in the nature of a stab in the dark, a matter of natural or instinctive belief rather than a matter of knowledge? No matter how enamoured we may be of common sense and our existing concepts we must admit that we do, in a definite way, go beyond the possible evidence, and so we still feel inclined to allow that although what the Idealist says is very odd and in the end unacceptable, nevertheless there is considerable justification for what he says. Why else is Berkeley of such historical importance? The Idealist may be strange but at least he's hard-headed. But things are very different when we turn to the Idealist's denial of motion (except for relative motion among items in the one percept). It is possible to deny that things move, that we bump into them and they into us, and so on, but this is very strange, and this time the Idealist has no defensive justification that he is being stricter, more hard-headed, taking less for granted, or anything of that sort. The Idealist's views about motion are simply to be described, and no doubt dismissed, as bizarre. Motion and action may not provide us with any special knowledge of the external world, but they are very important in our originally coming to conceive of what we perceive as an external world.

# LIST OF WORKS CITED

### ABBREVIATIONS

A     *Analysis*
APQ     *American Philosophical Quarterly*
JP     *Journal of Philosophy*
M     *Mind*
P     *Philosophy*
PAS     *Proceedings of the Aristotelean Society*
PASS     *Proceedings of the Aristotelean Society*, Supplementary Volumes
PBA     *Proceedings of the British Academy*
PR     *Philosophical Review*
PQ     *Philosophical Quarterly*

ANSCOMBE, G. E. M.
    (I) *Intention*, Blackwell, 1957.
ARMSTRONG, D. M.
    (I) *Perception and the Physical World*, Routledge and Kegan Paul, 1961.
    (II) *Bodily Sensations*, Routledge and Kegan Paul, 1963.
AUSTIN, J. L.
    (I) 'Other Minds', in *Philosophical Papers*, Oxford, 1961
    (II) *Sense and Sensibilia*, Oxford, 1962.
AYER, A. J.
    (I) *Language, Truth and Logic*, 2nd ed., Gollancz, 1946.
    (II) *Foundations of Empirical Knowledge*, Macmillan, 1940.
    (III) 'The Terminology of Sense Data', in *Philosophical Essays*, Macmillan, 1954.
    (IV) 'Phenomenalism', in *Philosophical Essays*.
    (V) 'Basic Propositions', in *Philosophical Essays*.
    (VI) *The Problem of Knowledge*, Penguin, 1956.
BERKELEY, G.
    (I) *Principles of Human Knowledge*, 1710, 2nd ed., 1734.
    (II) *Three Dialogues between Hylas and Philonous*, 1713.
BERLIN, I.
    (I) 'Empirical Propositions and Hypothetical Statements', M 1950.
BOUWSMA, O. K.
    (I) 'Moore's Theory of Sense Data', in *The Philosophy of G. E. Moore*, ed. Schilpp, Northwestern, 1942.
BRITTON, K. W.
    (I) 'Seeming', PASS 1952.
BROAD, C. D.
    (I) 'Is There "Knowledge by Acquaintance"?', PASS 1919.
    (II) *The Mind and Its Place in Nature*, Kegan Paul, 1925.
    (III) 'Berkeley's Argument about Material Substance', PBA 1942.

BROWN, N.
(I) 'Sense Data and Physical Objects', M 1957.
CHISHOLM, R.
(I) 'The Problem of the Speckled Hen', M 1942.
(II) 'The Theory of Appearing', in *Philosophical Analysis*, ed. Black, Cornell, 1950.
CLARK, M.
(I) 'Knowledge and Grounds', A 1963.
DUCASSE, C. J.
(I) 'Moore's Refutation of Idealism', in *The Philosophy of G. E. Moore*, ed. Schilpp, Northwestern, 1942.
GETTIER, E.
(I) 'Is Justified True Belief Knowledge?', A 1963.
GRICE, H. T.
(I) 'The Causal Theory of Perception', PASS 1961.
(II) 'Some Remarks on the Senses', in *Analytic Philosophy*, ed. Butler, Blackwell, 1962.
HALL, R.
(I) 'A Note on Sense Data', M 1964.
HAMLYN, D. W.
(I) *Sensation and Perception*, Routledge and Kegan Paul, 1961.
HAMPSHIRE, S.
(I) 'Identification and Existence', in *Contemporary British Philosophy III*, ed. Lewis, Allen and Unwin, 1956.
(II) 'Perception and Identification', PASS 1961.
(III) *Thought and Action*, Chatto and Windus, 1960.
HARDIE, W. F. R.
(I) 'The Paradox of Phenomenalism', PAS 1945-6.
HIRST, R. J.
(I) *The Problems of Perception*, Allen and Unwin, 1959.
(II) in *Human Senses and Perception*, ed. Wyburn, Oliver and Boyd, 1964.
HUME, D.
(I) *Treatise of Human Nature*, 1739 and 1740.
LEWIS, C. I.
(I) 'Professor Chisholm and Empiricism', JP 1948.
LOCKE, D. B.
(I) 'Strawson's Auditory Universe', PR 1961.
(II) 'The Privacy of Pains', A 1964.
(III) 'Intention and Intentional Action', forthcoming.
MOORE, G. E.
(I) 'The Refutation of Idealism', in *Philosophical Studies*, Kegan Paul, 1922.
(II) 'The Status of Sense Data', in *Philosophical Studies*.
(III) 'A Reply to My Critics', in *The Philosophy of G. E. Moore*, ed. Schilpp, Northwestern, 1942.
(IV) *Some Main Problems of Philosophy*, Allen and Unwin, 1953.

(V) 'Visual Sense Data', in *Philosophy in Mid-Century*, ed. Mace, Allen and Unwin, 1957.
(VI) 'A Defence of Common Sense', in *Philosophical Papers*, Allen and Unwin, 1959.
(VII) 'The Proof of an External World', in *Philosophical Papers*.
(VIII) *Common Place Book*, ed. Lewy, Allen and Unwin, 1962.

MUNDLE, C. K.
(I) 'Common Sense versus Mr Hirst's Theory of Perception', PAS 1959-60.

NELSON, J. O.
(I) 'An Examination of D. M. Armstrong's Theory of Perception', APQ 1964.

PAUL, G. A.
(I) 'Is There a Problem about Sense Data?', in *Logic and Language I*, ed. Flew, Blackwell, 1951.

PRICE, H. H.
(I) *Perception*, Methuen, 1932.
(II) *Hume's Theory of the External World*, Oxford, 1940.
(III) Review of Ayer's *Foundations of Empirical Knowledge*, M 1941.
(IV) 'Appearing', APQ 1964.

PRICHARD, H. A.
(I) 'Bertrand Russell on our Knowledge of the External World', in *Knowledge and Perception*, Oxford, 1950.
(II) 'The Sense Datum Fallacy', in *Knowledge and Perception*.
(III) 'Perception', in *Knowledge and Perception*.

QUINTON, A.
(I) 'The Problem of Perception', M 1955.

RUSSELL, B.
(I) *The Problems of Philosophy*, Hutchinson, 1st ed. Williams and Norgate, 1912.
(II) 'The Ultimate Constituents of Matter', in *Mysticism and Logic*, Allen and Unwin, 1917.
(III) 'Sense Data and Physics', in *Mysticism and Logic*.
(IV) *Analysis of Mind*, Allen and Unwin, 1922.
(V) *Outline of Philosophy*, Allen and Unwin, 1927
(VI) *Inquiry into Meaning and Truth*, Allen and Unwin, 1940.
(VII) 'A Reply to My Critics', in *The Philosophy of Bertrand Russell*, ed. Schilpp, Northwestern, 1944.

RYLE, G.
(I) *The Concept of Mind*, Hutchinson, 1949.
(II) *Dilemmas*, Cambridge, 1954.
(III) 'Sensations', in *Contemporary British Philosophy III*, ed. Lewis, Allen and Unwin, 1956.

SELLARS, W. F.
(I) 'Empiricism and the Philosophy of Mind', in *Science, Perception and Reality*, Routledge and Kegan Paul, 1963.

SIBLEY, F. N.
(I) 'Seeking, Scrutinising and Seeing', M 1955.

STACE, W. T.
  (I) 'The Refutation of Realism', in *Readings in Philosophical Analysis*, ed. Feigel and Sellars, Appleton-Century-Crofts, 1949.
STOUT, G. F.
  (I) 'Phenomenalism', PAS 1938-9.
STRAWSON, P. F.
  (I) *Individuals*, Methuen 1959.
  (II) 'Professor Ayer's *Problem of Knowledge*', P 1957.
  (III) 'Perception and Identification', PASS 1961.
URMSON, J. O.
  (I) *Philosophical Analysis*, Oxford, 1956.
WALTON, K.
  (I) 'The Dispensibility of Perceptual Inferences', M 1963.
WARNOCK, G. J.
  (I) *Berkeley*, Penguin, 1953.
  (II) 'Seeing', PAS 1954-5.
WHITE, A. R.
  (I) *G. E. Moore*, Blackwell, 1958.
  (II) 'The Causal Theory of Perception', PASS 1961.
  (III) *Attention*, Blackwell, 1964.
  (IV) 'The Alleged Ambiguity of "See" ', A 1963.
WISDOM, J.
  (I) *Other Minds*, Blackwell, 1952.
WITTGENSTEIN, L.
  (I) *Philosophical Investigations*, Blackwell, 1953.
ZEIDINS, R.
  (I) 'Conditions of Observation and States of Observers', PR 1956.

# INDEX

Achievement verbs, 28, 31–2
Action, 230–6
Alternative languages, 40–6
Anscombe, G. E. M., 232
Appearance, 37, 38, 92, 95, 103–12, 139–41, 165–6, 170–1, 183, 191–4, 197–201, 203, 218–21, 228
Appearance-determined qualities, 69–72, 87, 101, 105, 122, 189, 197, 200–1, 220–1
Armstrong, D. M., 17, 28–30, 57, 86, 115, 193, 195
Austin, J. L., 45, 95, 99, 105–7, 146, 161, 162, 188, 195–6, 227
A-veridical perception, 93
Ayer, A. J., 22, 39–46, 54–9, 65, 99, 101, 109, 126, 148, 154, 166, 173, 174, 190, 191, 193–5, 218, 223

Basis for knowledge, 206–10
Berkeley, G., 22, 36, 37, 51, 54, 58, 59, 69, 81–3, 109, 114, 115, 126, 130, 132, 136, 137, 176, 197, 236
Berlin, I., 57
Bouwsma, O. K., 168
Britton, K. W., 18
Broad, C. D., 13, 21, 79, 139, 182
Brown, N., 17

'Can', 51–4, 59–60
Causal processes, 113–8, 121
Causal theory, 22, 35–7, 45, 68, 77, 92, 113–20, 126, 129, 130, 132, 135–8, 141, 178
Chisholm, R., 166, 193
Claiming to know, 146–8
Clark, M., 151
Classification, 218–20
Colours, 69–72, 75–8, 98–103, 124, 188–92
Conceptual scheme, 131–4
Conditions of Observation, 69, 94, 99–102
Constancy and coherence, 137, 222–8
Construction of external world, 211–12, 228
Constructivism, 210–15

Delusion, 93–4, 105, 107

Derivation of knowledge, 158–9, 184–6, 196, 215–29
Descartes, R., 50, 211, 230
Description, and identification, 97–8, 110, 171, 201
Deutscher, M., 151
Dispositional properties, 123–4
Double vision, 108, 168
Dr Crippen, 53–4
Dreams, 17, 97

Egocentric Predicament, 127
Empiricism, 50, 145, 157–63, 206–11, 228, 230, 234
Epistemological priority, 206–10
Evidence, 146, 148, 150, 162, 229
Existence, 16–20, 50–9, 92–103, 166–7, 171, 174, 182–3
External objects, 20, 95–8, 120–5, *et passim*
Explanation, 58–9, 114, 126, 137
Extensive magnitude, 76–7

'Feel', 16
Foundations of knowledge, 157, 161–3, 184, 196, 204

Gettier, E., 151
Grice, H. T., 14, 38, 104

Hall, R., 21
Hallucinations, 17–20, 50–3, 87–9, 92–3, 95–8, 105, 107, 108, 111–12, 137, 138, 141, 166, 168, 174, 212–13, 216, 231
Hamlyn, D., 80
Hampshire, S., 132–3, 171, 178
Hardie, W., 58
Hirst, R. J., 18, 40–2, 79, 108, 114, 117, 139, 187
Hume, D., 222, 225
Hypotheticals, 51–4, 59–62

Idealism: 22, 35–7, 45, 47–50, 54, 56–9, 61, 68, 77, 80, 92, 107, 120, 124–38, 141, 152, 178, 224, 235–6
Identification, 62–4, 67, 75, 132–4
Identification, and description, 97–8, 110, 171, 174, 201, 205–6

Illusion, 37, 46, 92, 95, 105–12, 121, 165–6
Images, 17–20, 111, 168–9, 194
Immediate appearance, 191–2, 218–20, 228
Immediate perception, 47, 171–9
Incorrigibility, infallibility, 39, 86–7, 104, 162–3, 167, 186–92, 196, 197, 220–1
Indirect perception, 114, 175
Innate knowledge, 157, 205
Intensive magnitude, 76–7
Internal objects, 73–4, 78, 180–1
Introspective knowledge, 232, 233

Justification, 146–8; see also Right to be sure

Kant, I., 114, 222–6
Knowledge, 127–9, 145–60, 229, et Part II passim
Knowledge by acquaintance, 186–7
Knowledge of external objects (or reality), 113–14, 126–9, 138, 145, 152–63, 203–10, 212–17, 222–8, 230–6
Knowledge of perception-that, 196, 204–10, 214–22

Lewis, C. I., 66
Locke, J., 22, 36, 59, 69, 102, 126, 137, 236

Memory, 221–2
Mental—physical, 20, 181
Mill, J. S., 81
Moore, G. E., 21, 22, 38–40, 44, 57, 73, 80, 84, 129–31, 164–5, 168–72, 182, 186
Motion, 235–6
Muller-Lyer illusions, 92, 106–7
Mundle, C. K., 116

Nelson, J. O., 29
Neutral monism, 65, 136
Newton, I., 75
Non-observational knowledge, 232
Non-veridical perception, 92–5, 105–12, 137
Noticing, 30–2

Occam's Razor, 36–43, 136, 138

Pain, 82–91
Paradigm case argument, 130–1
Paul, G., 21, 38–42, 44
'Perceive', 15–20, 33
Perceiving-as, 32, 71, 89, 90, 94, 110, 120–5, 177, 191–2
Perception, 27–34, 202–3, et passim
Perception-that, 30, 32–4, 92–4, 172, 185–6, 192–6, 198, 205, 217
Percepts, 20–2, et passim
Performatives, 146
Perspectives, 139–41
Phenomenalism, 36, 42–3, 47–67, 120–2, 124–7, 135, 140, 167, 178, 214, 227, 230, 231
Physical objects, 20, 35–6, 40, 46, 47, 49, 74–5, 77, 95–8, 124, 129–32, 139–40
Physical–mental, 20, 181
Price, H. H., 21, 22, 39, 99, 108, 139, 174, 187, 188, 193, 194, 198, 223
Primary qualities, 68–78, 122
Pritchard, H. A., 140, 187–8
Privacy, 20, 37, 79, 85–6, 139–41, 181–2, 194–5, 197
Private language argument, 56
Proper objects, 72–4

Quinton, A. N., 191

'Real', 19–20, 95–103
Realism, 22, 35–8, 40–50, 57–9, 96, 102, 107–9, 117–20, 126–38, 140–1, 145, 152, 153, 178–9, 181–4, 187, 202
Reductionism, 210–15
Representative theory, 35, 113, 115, 117, 118
Right to be sure, 148–50, et Part II passim
Russell, B., 13, 21, 116–8, 138–41
Ryle, G., 28, 31, 34, 78, 80–1, 180–1

Scepticism, 133, 152–7, 161, 167, 211, 213, 221
Secondary qualities, 68–78, 122
Seeming, 105
Sellars, W., 161
Sensations, 78–91, 180
Sense data, 21–3, 164–203, 221, et passim
Sense datum knowledge, 162–3, 184–92, 203–10, 212–14, 216–22

# INDEX

Sense datum statements, 47–50, 55–67, 170, 177–8, 184, 199–201
Sense knowledge, 158–60, 204–36
Sense objects, 175–7, 221
Sensibilia, 35, 138–41
Sensory awareness, 27–30, 32, 79–81, 92–4, 172
Shape, 70–2, 101–2
Sibley, F., 27, 31
Solipsism, 48, 127, 136, 138, 223–36
Sounds, 71, 75–6, 90, 121–3
Spatial location, 52–3, 62–5, 139–41, 181
Stace, W. T., 80, 126
Stout, G. F., 58
Strawson, P. F., 62, 132–4, 149, 153, 178, 223–6
Subliminal knowledge, 232–5

Tautologous objects, 72–4, 176
Temperature, 76, 82–3
Theories of perception, 13–14, 35–46, 135–41, 164–5, 167
Time lag argument, 118

Urmson, J. O., 56

Veridical perception, 160

Walton, K., 173
Warnock, G. J., 34, 115, 173, 175, 176, 195, 227–8
White, A. R., 15, 31, 38, 83, 97–8, 168, 201
Wisdom, J. T., 63, 192
Wittgenstein, L., 187, 233

Zeidins, R., 100, 101

For Product Safety Concerns and Information please contact our EU representative GPSR@taylorandfrancis.com
Taylor & Francis Verlag GmbH, Kaufingerstraße 24, 80331 München, Germany

www.ingramcontent.com/pod-product-compliance
Lightning Source LLC
Chambersburg PA
CBHW070723020526
44116CB00031B/1459